Spending Quality Time with God
15 Minutes That Can Change Your Life

1. ***Relax*** (1 minute)
   - Take a minute to focus your mind on God.
   - Be still and know that I am God." Psalm 46:10a

2. ***Read*** the Word (4 minutes)
   - As you read, focus on listening to God as He speaks to you.
   - Have an open mind and an open heart to God's Word.

3. ***Reflect*** on the Word (3 minutes)
   - Think through what you just read. How does this scripture apply to my life?
   - How does this apply to my relationship with God, with my family, with the church, with others, and with myself?

4. ***Record*** (3 minutes)
   - Write down important questions, observations and personal applications from your reading.
   - Decide on one action step to take based on what you read today.

5. ***Request*** (4 minutes)
   - Pray to God with thanksgiving and express your needs to Him.
   - Pray for strength to follow through with your action step for the day.

This book is dedicated to all our families for their support, and to all those individuals who desire to grow in passion and zeal to Him!

We would like to acknowledge Phillip Johnson for his energy and dedication to make this book a reality, and Brad Harrub for his diligence to see it through to completion. Also, we thank the supporters of Focus Press who helped make this book a reality.

iGrow

Published by Focus Press, Inc.

Copyright © 2011 Focus Press, Inc. and Phillip Johnson
International Standard Book Number 978-0-937729-3-4

First Printing   2011
Second Printing 2016

Cover by: Nick Long
Cover image: © Thinkstock

Printed in the United States of America

ALL RIGHTS RESERVED

Focus Press, Inc.
625 Bakers Bridge Ave. , Ste. 105
Franklin, TN 37067
www.focuspress.org

Date	Lesson	Date	Lesson	Date	Lesson
Jan 01	1	Feb 06	37	Mar 14	73
Jan 02	2	Feb 07	38	Mar 15	74
Jan 03	3	Feb 08	39	Mar 16	75
Jan 04	4	Feb 09	40	Mar 17	76
Jan 05	5	Feb 10	41	Mar 18	77
Jan 06	6	Feb 11	42	Mar 19	78
Jan 07	7	Feb 12	43	Mar 20	79
Jan 08	8	Feb 13	44	Mar 21	80
Jan 09	9	Feb 14	45	Mar 22	81
Jan 10	10	Feb 15	46	Mar 23	82
Jan 11	11	Feb 16	47	Mar 24	83
Jan 12	12	Feb 17	48	Mar 25	84
Jan 13	13	Feb 18	49	Mar 26	85
Jan 14	14	Feb 19	50	Mar 27	86
Jan 15	15	Feb 20	51	Mar 28	87
Jan 16	16	Feb 21	52	Mar 29	88
Jan 17	17	Feb 22	53	Mar 30	89
Jan 18	18	Feb 23	54	Mar 31	90
Jan 19	19	Feb 24	55	Apr 01	91
Jan 20	20	Feb 25	56	Apr 02	92
Jan 21	21	Feb 26	57	Apr 03	93
Jan 22	22	Feb 27	58	Apr 04	94
Jan 23	23	Feb 28	59	Apr 05	95
Jan 24	24	Mar 01	60	Apr 06	96
Jan 25	25	Mar 02	61	Apr 07	97
Jan 26	26	Mar 03	62	Apr 08	98
Jan 27	27	Mar 04	63	Apr 09	99
Jan 28	28	Mar 05	64	Apr 10	100
Jan 29	29	Mar 06	65	Apr 11	101
Jan 30	30	Mar 07	66	Apr 12	102
Jan 31	31	Mar 08	67	Apr 13	103
Feb 01	32	Mar 09	68	Apr 14	104
Feb 02	33	Mar 10	69	Apr 15	105
Feb 03	34	Mar 11	70	Apr 16	106
Feb 04	35	Mar 12	71	Apr 17	107
Feb 05	36	Mar 13	72	Apr 18	108

Date	Lesson	Date	Lesson	Date	Lesson
Apr 19	109	May 25	145	Jun 30	181
Apr 20	110	May 26	146	Jul 01	182
Apr 21	111	May 27	147	Jul 02	183
Apr 22	112	May 28	148	Jul 03	184
Apr 23	113	May 29	149	Jul 04	185
Apr 24	114	May 30	150	Jul 05	186
Apr 25	115	May 31	151	Jul 06	187
Apr 26	116	Jun 01	152	Jul 07	188
Apr 27	117	Jun 02	153	Jul 08	189
Apr 28	118	Jun 03	154	Jul 09	190
Apr 29	119	Jun 04	155	Jul 10	191
Apr 30	120	Jun 05	156	Jul 11	192
May 01	121	Jun 06	157	Jul 12	193
May 02	122	Jun 07	158	Jul 13	194
May 03	123	Jun 08	159	Jul 14	195
May 04	124	Jun 09	160	Jul 15	196
May 05	125	Jun 10	161	Jul 16	197
May 06	126	Jun 11	162	Jul 17	198
May 07	127	Jun 12	163	Jul 18	199
May 08	128	Jun 13	164	Jul 19	200
May 09	129	Jun 14	165	Jul 20	201
May 10	130	Jun 15	166	Jul 21	202
May 11	131	Jun 16	167	Jul 22	203
May 12	132	Jun 17	168	Jul 23	204
May 13	133	Jun 18	169	Jul 24	205
May 14	134	Jun 19	170	Jul 25	206
May 15	135	Jun 20	171	Jul 26	207
May 16	136	Jun 21	172	Jul 27	208
May 17	137	Jun 22	173	Jul 28	209
May 18	138	Jun 23	174	Jul 29	210
May 19	139	Jun 24	175	Jul 30	211
May 20	140	Jun 25	176	Jul 31	212
May 21	141	Jun 26	177	Aug 01	213
May 22	142	Jun 27	178	Aug 02	214
May 23	143	Jun 28	179	Aug 03	215
May 24	144	Jun 29	180	Aug 04	216

Date	Lesson	Date	Lesson	Date	Lesson
Aug 05	217	Sep 10	253	Oct 16	289
Aug 06	218	Sep 11	254	Oct 17	290
Aug 07	219	Sep 12	255	Oct 18	291
Aug 08	220	Sep 13	256	Oct 19	292
Aug 09	221	Sep 14	257	Oct 20	293
Aug 10	222	Sep 15	258	Oct 21	294
Aug 11	223	Sep 16	259	Oct 22	295
Aug 12	224	Sep 17	260	Oct 23	296
Aug 13	225	Sep 18	261	Oct 24	297
Aug 14	226	Sep 19	262	Oct 25	298
Aug 15	227	Sep 20	263	Oct 26	299
Aug 16	228	Sep 21	264	Oct 27	300
Aug 17	229	Sep 22	265	Oct 28	301
Aug 18	230	Sep 23	266	Oct 29	302
Aug 19	231	Sep 24	267	Oct 30	303
Aug 20	232	Sep 25	268	Oct 31	304
Aug 21	233	Sep 26	269	Nov 01	305
Aug 22	234	Sep 27	270	Nov 02	306
Aug 23	235	Sep 28	271	Nov 03	307
Aug 24	236	Sep 29	272	Nov 04	308
Aug 25	237	Sep 30	273	Nov 05	309
Aug 26	238	Oct 01	274	Nov 06	310
Aug 27	239	Oct 02	275	Nov 07	311
Aug 28	240	Oct 03	276	Nov 08	312
Aug 29	241	Oct 04	277	Nov 09	313
Aug 30	242	Oct 05	278	Nov 10	314
Aug 31	243	Oct 06	279	Nov 11	315
Sep 01	244	Oct 07	280	Nov 12	316
Sep 02	245	Oct 08	281	Nov 13	317
Sep 03	246	Oct 09	282	Nov 14	318
Sep 04	247	Oct 10	283	Nov 15	319
Sep 05	248	Oct 11	284	Nov 16	320
Sep 06	249	Oct 12	285	Nov 17	321
Sep 07	250	Oct 13	286	Nov 18	322
Sep 08	251	Oct 14	287	Nov 19	323
Sep 09	252	Oct 15	288	Nov 20	324

Date	Lesson
Nov 21	325
Nov 22	326
Nov 23	327
Nov 24	328
Nov 25	329
Nov 26	330
Nov 27	331
Nov 28	332
Nov 29	333
Nov 30	334
Dec 01	335
Dec 02	336
Dec 03	337
Dec 04	338
Dec 05	339
Dec 06	340
Dec 07	341
Dec 08	342
Dec 09	343
Dec 10	344
Dec 11	345
Dec 12	346
Dec 13	347
Dec 14	348
Dec 15	349
Dec 16	350
Dec 17	351
Dec 18	352
Dec 19	353
Dec 20	354
Dec 21	355
Dec 22	356
Dec 23	357
Dec 24	358
Dec 25	359
Dec 26	360

Date	Lesson
Dec 27	361
Dec 28	362
Dec 29	363
Dec 30	364
Dec 31	365

Lesson 1

Relax
-Spend one minute thinking of two blessings God has given you.

Read
-Devote four minutes to reading Matthew 1.

Reflect
-Take three minutes to think through the following questions.
1. Which verse impacted you the most from today's reading?
2. Would you say you've become more Christ-like or less Christ-like over the last year? Why?
3. How would you describe Mary based on what you read about her today?
4. What did you learn from Joseph in Matthew 1 that will benefit you in your family relationships?

Record
-Spend three minutes answering the following questions.
1. Write down one question or observation you have over today's reading. _____
   _____

2. How will today's reading help you in your relationship with Jesus? _____
   _____

3. Write down one action step you will take based on your reading from Matthew 1. *I will* _____
   _____

Request
-Devote four minutes to talking with God.

Lesson 2

Relax
  -Spend one minute thinking of the majesty and beauty of heaven. In your opinion, what will be the greatest thing about heaven? _____

Read
  -Devote four minutes reading Psalm 1 through Psalm 4.

Reflect
  -Take three minutes to think through the following questions.
  1. What was your favorite verse from today's reading?
  2. How does David distinguish between the righteous and the wicked in Psalm 1?
  3. What can we learn about talking to God from David in these Psalms?

Record
  -Spend three minutes answering the following questions.
  1. What are two characteristics you learned about God from these Psalms? _____

  2. How will this reading help you build a stronger relationship with God? _____

  3. How can these Psalms help you in your interactions with those who are wicked? _____

  4. Write out one action step you will take based on today's reading. *I will* _____

Request
  -Take four minutes to talk with God.

Lesson 3

Relax
-Spend one minute thinking of your favorite aspect of God's creation.

Read
-Spend four minutes reading Matthew 2. *(You may be able to read it through two times)*

Reflect
-Take three minutes to think through the following questions.
1. If you knew you had only five days left to live, how would you spend those five days?
2. *Why don't you do these things now?*
3. How did God work in the lives of people in Matthew 2?
4. How does God work in the lives of people today?
5. How would you describe the faith of Joseph?

Record
-Devote three minutes to answering the following questions.
1. How can these verses help you in your relationships with your family? _____

2. How will these verses help you build a stronger relationship with Jesus?_____

3. Write out an action step you will take in your life based on your reading today. *I will* _____

Request
-Take four minutes to talk with God.

Lesson 4

Relax
-Spend a minute thinking of the words of Psalm 23:1, *"The Lord is my shepherd I shall not want."*

Read
-Devote four minutes to reading Psalm 5, 6 and 7.

Reflect
-Take three minutes to think through the following questions.
1. What has been your biggest accomplishment in life so far?
2. How would you describe David's relationship with God?
3. How would you describe your relationship with God?
4. What verse(s) grabbed your attention in today's reading?

Record
-Spend three minutes answering the following questions.
1. Write down one question or observation you have over these Psalms. _____
_____

2. What is one thing you learned from David that can help you in your daily walk with Christ? _____
_____

3. Write out one action step you will take in your life based on today's Bible reading. *I will* _____
_____

Request
-Devote four minutes to talking with God in prayer.

Lesson 5

Relax
    -Spend one minute reflecting on one struggle God has delivered you through.

Read
    -Spend four minutes reading through Matthew 3.

Reflect
    -Take three minutes to think through the following questions.
1. What verse grabbed your attention in Matthew 3?
2. Why was Jesus baptized by John even though He lived a sinless life? (see verse 15)
3. Would you describe John as a humble person based on what you read in this chapter?

Record
    -Devote three minutes to answering the following questions.
1. What amazed you the most about Christ from your reading? _____
2. Write out one question you have over today's reading in Matthew 3. _____
3. Describe God's feelings towards Jesus. _____
4. How can this chapter help you in your relationship with God? _____
5. Write down one action step you will take in your life based on your reading today. *I will* _____

Request
    -Spend four minutes talking with God.

Lesson 6

Relax
-Spend one minute thinking about the following question: *What characteristic of God brings you the most peace in your life?*

Read
-Devote four minutes to reading Psalm 8 and Psalm 9.

Reflect
-Take three minutes to think through the following questions.
1. How would you describe David's mood in these two Psalms?
2. If you had one minute on national television to say something about God, what would you say?
3. Which verse(s) grabbed your attention today?

Record
-Spend three minutes answering the following questions.
1. List one important aspect of prayer we learn from David in these two Psalms. _____

2. Write down one question or observation from Psalm 8 and 9. _____

3. How can these Psalms help you in your relationship with God? _____

4. Write down one action step you will take based on today's reading. *I will* _____

Request
-Devote four minutes to prayer.

Lesson 7

Relax
-Spend one minute with your eyes closed thinking of Jesus sitting at the right hand of God in heaven.

Read
-Devote four minutes to reading through Matthew 4.

Reflect
-Take three minutes to think through the following questions.
1. How did Jesus end up in the wilderness?
2. In your opinion, which of these temptations was the most difficult for Jesus to overcome? Why?
3. How would you describe Satan based on your reading today?
4. What did you learn from Matthew 4 that will help you overcome Satan?

Record
-Spend three minutes answering the following questions.
1. List two characteristics you learned about Jesus from this chapter. _____
2. What did you learn about Jesus' purpose for life from Matthew 4:23? _____
3. Write down one personal observation or question from Matthew 4. _____
4. Write out one action step you will take based on today's reading. *I will* _____

Request
-Spend four minutes talking to God.

Lesson 8

Relax
-Spend one minute meditating on the words of Philippians 4:13. *"I can do all things through Christ who strengthens me."*

Read
-Devote four minute to reading Psalm 10 and 11.

Reflect
-Take three minutes to think through the following questions.
1. What is David praying for in these Psalms?
2. On a scale of 1-10, with 10 being powerful how would you rate your prayer life?
3. If Jesus were to physically follow you around one day, would you do anything differently?

Record
-Spend three minutes answering the following questions.
1. Write down one question or personal observation you have over these Psalms. _____

2. David is constantly asking God questions. What is one question you would like to ask God? _____

3. How can these Psalms help you build a stronger relationship with God? _____

4. Write out one action step you will take based on today's reading. *I will* _____

Request
-Take four minutes to talk with God.

Lesson 9

Relax
-Spend one minute reflecting on one of your greatest personal achievements.

Read
-Devote four minutes to reading Matthew 5.

Reflect
-Take three minutes to think through the following questions.
1. Which one of the beatitudes is most difficult for you to live out in your life?
2. Who's the most spiritual person you know? Why?
3. If someone asked if you knew you were going to heaven, what would you say?
4. How would you describe the teachings of Jesus?

Record
-Spend three minutes answering the following questions.
1. Write down one question or observation you have over this chapter. _____

2. How will this chapter help you in your relationships with your family? _____

3. How will Matthew 5 assist you in building a stronger love for Jesus? _____

4. List one action step you will take in your life based on today's reading. *I will* _____

Request
-Devote four minutes to praying to God.

Lesson 10

Relax
 -Spend one minute thinking on the words of Hebrews 13:8. *"Jesus Christ is the same yesterday, today and forever."*

Read
 -Devote four minutes to reading Psalm 12 – Psalm 15.

Reflect
 -Take three minutes to think through the following questions.
  1. How would you describe the words of the Lord?
  2. Do you think God is happy with the way He's represented in the world?
  3. What message are you telling the world about God based on how you are currently living?

Record
 -Spend three minutes answering the following questions.
  1. Write out one question or observation you have from today's reading. _____

  2. What are some characteristics we learn about God from these Psalms? _____

  3. How can these Psalms help you in your relationship with God? _____

  4. Write out one action step you will take in your life based on today's reading. *I will* _____

Request
 -Devote four minutes to talking with God.

Lesson 11

Relax
    -Take a minute to close your eyes and reflect on the best gift you have ever received.

Read
    -Devote four minutes to reading Matthew 6.

Reflect
    -Take three minutes to think through the following questions.
1. Why did Jesus teach His disciples how to pray?
2. Think about one way you can improve your prayer life.
3. Which verse(s) seemed as if Jesus knew your struggles and was talking directly to you?
4. Of all the problems of the world, what problem do you think saddens God the most?

Record
    -Spend three minutes answering the following questions.
1. List two examples of hypocritical behavior described by Jesus. _____
2. List one thing you constantly worry about. _____
3. What is Jesus' solution to worry and anxiety? _____ _____
4. Write out one question or observation you have over Matthew 6. _____ _____
5. Write down one action step you will take in your life based on today's reading. *I will* _____ _____

Request
    -Devote four minutes to prayer.

Lesson 12

Relax
   -Take a minute to think on these words. *Out of all the majestic things God has created, you are the most important in the eyes of God.*

Read
   -Devote four minutes to reading Psalm 16 and Psalm 17.

Reflect
   -Take three minutes to think through the following questions.
   1. In Psalm 7, David says he will bless the Lord. How can you bless the Lord in your life?
   2. What is your favorite verse from today's reading?
   3. In you opinion, what is the biggest obstacle that keeps people from believing God?

Record
   -Spend three minutes answering the following questions.
   1. Write down one question or observation you have over today's reading. _____
   2. List two things David says God will provide for him. _____
   3. What is one request David made to God in Psalm 17? _____
   4. Write down the name of one person you need to include in your prayers. _____
   5. Write down one action step you will take in your life based on today's reading. *I will* _____

Request
   -Devote four minutes to prayer.

Lesson 13

Relax
-Spend one minute talking to God about one problem you are currently facing.

Read
-Devote four minutes to reading Matthew 7.

Reflect
-Take three minutes to think through the following questions.
1. What verse from Matthew 7 brings you the most comfort?
2. Is it OK in the eyes of God to pick and choose which commands you follow?
3. What did you learn about prayer from Matthew 7?

Record
-Spend three minutes answering the following questions.
1. List one way this chapter will help you build a better relationship with Jesus. _____
2. Write down one teaching Jesus shares in this chapter concerning heaven. _____
3. What is the one major difference between the wise man and the foolish man? _____
4. How can you be sure you are on the path to heaven? _____
5. Write down one action step you will take in your life based on today's reading. *I will* _____

Request
-Spend four minutes talking with God.

Lesson 14

Relax
    -Spend one minute meditating on the words of Hebrews 13:5. *"I (Jesus) will never leave you nor forsake you."*

Read
    -Devote four minutes to reading through Psalm 18.

Reflect
    -Take three minutes to think about the following questions.
1. Which verse from Psalm 18 made the biggest impact on you?
2. David writes this Psalm after being delivered from the hand of Saul. *Reflect on one time when God delivered you through a difficult time in your life.*
3. What is one new thing you learned about God from today's reading?
4. David's prayers were balanced between making requests to God, and saying thanks to God. *Would you describe your prayers as being balanced?*

Record
    -Spend three minutes answering the following questions.
1. List one way this chapter will help you in your relationship with God. _____
2. Write down one question or observation you have over Psalm 18. _____
3. Write down one action step you will take in your life based on what you learned from Psalm 18. *I will* __

Request
    -Devote four minutes to prayer.

Lesson 15

Relax
-Spend one minute meditating on the words of your favorite worship song.

Read
-Devote four minutes to reading Matthew 8.

Reflect
-Take three minutes to meditate on the following questions.
1. How can Matthew 8 help you make a difference in the lives of others?
2. If you could meet anybody from this chapter, whom would you choose to meet? Why?
3. Which verse(s) grabbed your attention today?
4. How would you describe Jesus based on what you learned about Him from this chapter?

Record
-Spend three minutes answering the following questions.
1. How can this chapter help you build a stronger relationship with Jesus? _____
_____
2. What is the most significant lesson you learned today? _____
3. Write down one question or observation you have over Matthew 8. _____
_____
4. Write down one action step you will take in your life based on what you learned today. *I will* _____
_____

Request
-Devote four minutes to prayer.

Lesson 16

Relax
-Spend one minute meditating on the words of your favorite verse from God's word.

Read
-Devote four minutes to reading Psalm 19 through Psalm 21.

Reflect
-Take three minutes to think through the following questions.
1. If you could ask God any question, what question would you ask?
2. Where do you feel closest to God?
3. How would you describe the words of God?
4. Why does David describe God's words as more valuable than gold?

Record
-Take three minutes to answer the following questions.
1. How can this chapter help you build a stronger relationship with God? _____

2. Write down one question or observation from your reading today. _____

3. Write down one action step you will take in your life based on today's reading from Psalms. *I will* _____

Request
-Devote four minutes to talking with God.

Lesson 17

Relax
  -Take one minute to meditate on a time in your life when you were faithful to God.

Read
  -Spend four minutes reading through Matthew 9.

Reflect
  -Take three minutes to think through the following questions.
  1. Which person demonstrated the most faith in this chapter?
  2. How would you describe your faith?
  3. What did Jesus mean by what He said in verse 12?
  4. Is Jesus' statement in verse 37 applicable to today's world?

Record
  -Devote three minutes to answering the following questions.
  1. List one new characteristic you learned about Jesus from this chapter. _____

  2. Write down one question or observation you have over Matthew 9. _____

  3. How will this chapter help you build a stronger relationship with Jesus? _____

  4. Write down one action step you will take in your life based on today's reading. *I will* _____

Request
  -Take four minutes to talk with God.

Lesson 18

Relax
-Take a minute to reflect on the events of what you consider to be the best day of your life.

Read
-Spend four minutes reading Psalm 22.

Reflect
-Take three minutes to think through the following questions.
1. What changed David's mood from verse 1 to verse 24?
2. What is one new characteristic you learned about God today?
3. Who do you know that is a good representative of God's character?
4. Which verse in Psalm 22 shows abortion is wrong in the eyes of God?

Record
-Devote three minutes to answering the following questions.
1. When did your relationship with God begin? _____
2. How would you describe your relationship with God right now? _____
3. How can this chapter help you build a stronger bond with God? _____
4. Write out one action step you will take in your life based on today's reading. *I will* _____

Request
-Spend four minutes talking with God.

Lesson 19

Relax
-Take one minute to meditate on the words of Romans 8:1. *"Therefore, there is now no condemnation for those who are in Christ Jesus."*

Read
-Spend four minutes reading Matthew 10.

Reflect
-Take three minutes to think through the following questions.
1. Does it cost to follow Jesus? *What are some costs involved in following Jesus?*
2. How can this chapter help you make a difference in the lives of others?
3. What thoughts would be going through your mind if you would have been one of Jesus' disciples?

Record
-Devote three minutes to answering the following question.
1. List one way this chapter will help you build a stronger relationship with Jesus. _____

2. Write down one question or observation you have over Matthew 10. _____

3. What are you doing to fulfill the command given by Jesus in verse 32? _____
4. Write out one action step you will take in your life based on your reading today. *I will* _____

Request
-Spend four minutes talking with God.

Lesson 20

Relax
-Spend one minute talking with God about the events of your day.

Read
-Devote four minutes to reading through Psalm 23. *(You may have time to memorize this comforting Psalm)*

Reflect
-Take three minutes to think through the following questions.
1. What did you learn from this Psalm that will help your daily service?
2. Which verse(s) in this Psalm brings you the most comfort?
3. What are some things people want for today?
4. Is the Lord your shepherd or are you constantly wanting?

Record
-Spend three minutes answering the following questions.
1. Write down one question or observation you have over Psalm 23. _____

2. List one thing you learned from this Psalm that will help you build a stronger relationship with God. ___

3. Write down one action step you will take in your life based on your reading today. *I will* _____

Request
-Devote four minutes to talking with God.

Lesson 21

Relax
-Spend one minute meditating on a time when you felt God's presence in your life.

Read
-Devote four minutes to reading Matthew 11.

Reflect
-Take three minutes to think through the following questions.
1. How will this chapter help you in your personal walk with Jesus?
2. What did you learn today that will help you reach out to your friends, neighbors and relatives?
3. What impact will Matthew 11 have on your relationship with your family?
4. Why did Jesus perform miracles?

Record
-Spend three minutes answering the following questions.
1. Write down one question or observation you have over Matthew 11. _____
_____
2. Which verse(s) grabbed your attention from today's reading? _____
3. Write down one action step you will take in your life based on today's reading. *I will* _____
_____

Request
-Devote four minutes to talking with God.

Lesson 22

Relax

-Take a minute to personalize John 3:16. *"For God so loved **me** that He gave His only begotten Son that whoever believes in Him shall not perish, but have eternal life."*

Read

-Devote four minutes to reading Psalm 25.

Reflect

-Take three minutes to think through the following questions.
1. How often have you prayed a prayer like David's in verses 4 and 5?
2. How would your life change if you always allowed God to guide you?
3. What does it look like to hope in God all day long?
4. How can this chapter help you influence those who don't have a relationship with God?

Record

-Spend three minutes answering the following questions.
1. How can this chapter help you in your relationship with God? _____
2. Write down one question or observation you have over Psalm 25. _____
3. List one action step you will take in your life based on today's reading. *I will* _____

Request

-Devote four minutes to talking with God remembering to ask for forgiveness and strength.

Lesson 23

Relax
    -Spend one minute meditating on how good it feels to be forgiven of some wrong.

Read
    -Devote four minutes to reading Matthew 12.

Reflect
    -Take three minutes to think through the following questions.
1. What does the statement 'Jesus is Lord' mean for you in your life?
2. How do you think Jesus felt when accused of being a follower of Satan?
3. How can this chapter help you make an impact on your family and friends?

Record
    -Spend three minutes answering the following questions.
1. What is the importance of the words we speak as described by Jesus in verse 37? _____

2. List one thing that surprised you about Jesus from this chapter. _____

3. Write down one question or observation you have over Matthew 12. _____

4. List one action step you will take in your life based on your reading today. *I will* _____

Request
    -Devote four minutes to talking with God.

Lesson 24

Relax
-Spend one minute thinking of the words God said to Joshua in Joshua 1:8, *"Be strong and courageous! Do not tremble or be dismayed, for the Lord your God is with you wherever you go."*

Read
-Devote four minutes to reading Psalm 26 and Psalm 27.

Reflect
-Take three minutes to think through the following questions.
1. How confident are you about God's presence being with you at all times?
2. Why do you think the things David sought after were so important to him?
3. How can these Psalms help you in your daily service to others?

Record
-Spend three minutes answering the following questions.
1. List one thing from these Psalms that will help you build a stronger relationship with God. _____

2. Write down one question or observation you have over these two Psalms. _____

3. List one action step you will take in your life based on today's reading. *I will* _____

Request
-Devote four minutes to talking with God.

Lesson 25

Relax
-Spend one minute meditating on how great the kingdom of heaven will be.

Read
-Devote four minutes to reading through Matthew 13.

Reflect
-Take three minutes to think through the following questions.
1. Which parable from this chapter impacted you the most?
2. How can this chapter help you reach out to those who are lost?
3. What is the most important lesson you learned from the parable of the sower?
4. Which type of soil in the parable best represents the condition of your heart right now?

Record
-Spend three minutes answering the following questions.
1. List one way this chapter will help you strengthen your relationship with Jesus. _____
   _____
2. Write down one question or observation you have over Matthew 13. _____
   _____
3. List one action step you will take in your life based on today's reading. *I will* _____
   _____

Request
-Devote four minutes to talking with God.

Lesson 26

Relax
-Spend one minute in a quiet place talking with God.

Read
-Devote four minutes to reading Psalm 28 through Psalm 30.

Reflect
-Take three minutes to think through the following questions.
1. What was the focus of David's prayer in Psalm 28?
2. What was the focus of his prayer in Psalm 30?
3. Reflect on a time in your life when God answered your cry for deliverance.
4. Read Psalm 30:5—What do you learn about life from this verse?

Record
-Spend three minutes answering the following questions.
1. Write down one question or observation from your reading today. _____

2. List one way these Psalms will help you build a better relationship with God. _____

3. How can these Psalms help you as you try to serve others? _____

4. Write out one action step you will take in your life based on your reading today. *I will* _____

Request
-Devote four minutes to talking with God.

Lesson 27

Relax
-Spend one minute meditating on your favorite story of Jesus.

Read
-Devote four minutes to reading Matthew 14.

Reflect
-Take three minutes to think through the following questions.
1. Why was John the Baptist beheaded?
2. What are some consequences of sinfulness?
3. What amazed you the most about Jesus in this chapter?
4. How would you describe Jesus' feelings for other people? *Do you have this kind of love for others?*
5. Is Jesus' question in verse 31 applicable to your life?

Record
-Spend three minutes answering the following questions.
1. How can this chapter in Matthew help you in your daily ministry? _____

2. Write down one question or observation you have over Matthew 14. _____

3. List one positive and negative thing you learned from Peter in this chapter. _____

4. List one action step you will take in your life based on today's reading. *I will* _____

Request
-Devote four minutes to talking with God.

Lesson 28

Relax
 -Spend one minute meditating on the words found in Psalm 31:24.

Read
 -Devote four minutes to reading through Psalm 31.

Reflect
 -Take three minutes to think through the following questions.
 1. Why do you think the enemy is allowed to have influence in the world?
 2. Have you ever felt as if no one could understand what you were going through?
 3. What comforting thought did you learn about God from David's statement in verse 7?

Record
 -Spend three minutes answering the following questions.
 1. Write down one question or observation you have over Psalm 31. _____

 2. List one way this chapter will help you in your personal relationship with God. _____

 3. How can this chapter help you reach out to other people who are going through tough times? _____

 4. List one action step you will take in your life based on today's reading. *I will* _____

Request
 -Devote four minutes to talking with God.

Lesson 29

Relax
-Spend one minute thinking on how wonderful it will be to finally meet Jesus face to face.

Read
-Devote four minutes to reading Matthew 15.

Reflect
-Take three minutes to think through the following questions.
1. Why is it difficult to overcome 'traditions' and simply follow God's Word?
2. How important is the condition of our hearts?
3. What condition is your spiritual heart in right now?
4. How did Jesus tell us we can judge the condition of our hearts?

Record
-Spend three minutes answering the following question.
1. List one way this chapter will help you strengthen your relationship with your Lord. _____

2. How can this chapter assist you in reaching out to the lost of this community? _____

3. Write down one question or observation you have over Matthew 15. _____

4. List one action step you will take in your life based on today's reading. *I will* _____

Request
-Devote four minutes to talking with God.

Lesson 30

Relax
-Spend one minute thanking God for the blessings He has given you in your life.

Read
-Devote four minutes to reading Psalm 32.

Reflect
-Take three minutes to think through the following questions.
1. What is the greatest blessing one has in Jesus Christ?
2. How do you feel when someone forgives you of some wrong?
3. Would you describe yourself as a forgiving person?
4. How wonderful is the feeling of being forgiven?

Record
-Spend three minutes answering the following questions.
1. How can this chapter help you improve your relationship with God? _____

2. List one thing you learned today that will help you in your ministry to other people. _____

3. Write down one question or observation you have from Psalm 32. _____

4. List one action step you will take in your life based on today's reading. *I will* _____

Request
-Devote four minutes to talking with God.

Lesson 31

Relax
-Spend one minute meditating on the following question.
Who is Jesus?

Read
-Devote four minutes to reading through Matthew 16.

Reflect
-Take three minutes to think through the following questions.
1. Why were people confused as to who Jesus really was?
2. Is there still confusion in the world today concerning Jesus? Why?
3. Why was Peter rebuked by Jesus in verse 23?
4. Reflect on a time in your life when your plans were different than God's plan.

Record
-Spend three minutes answering the following questions.
1. Write down one question or observation you have over Matthew 16. _____
_____
2. List one way this chapter can help you improve your relationship with Jesus. _____
_____
3. List something new you learned from today's reading. _____
4. Write out one action step you will take in your life based on your reading today. *I will* _____
_____

Request
-Devote four minutes to talking with God.

Lesson 32

Relax
-Spend one minute meditating on the words of a favorite worship song.

Read
-Devote four minutes to reading Psalm 33.

Reflect
-Take three minutes to think through the following questions.
1. If you were asked to describe God, what would you say about Him?
2. Read verse 12: Would you describe our nation as one whose God is the Lord? Why or why not?
3. Why has it seemed that our nation has drifted away from loving God?

Record
-Spend three minutes answering the following questions.
1. List one suggestion you would offer this nation that would help us to turn back to God. _____
2. Write down one question or observation you have over Psalm 33. _____
3. What did you learn about God in this Psalm that will strengthen your personal relationship with Him? ___
4. List one action step you will take in your life based on Psalm 33. *I will* _____

Request
-Devote four minutes to talking with God.

Lesson 33

Relax
-Spend one minute meditating on this question. *In what ways have you been blessed by walking with God?*

Read
-Devote four minutes to reading Matthew 17.

Reflect
-Take three minutes to think through the following questions.
1. What is the most important lesson you learned from Matthew 17?
2. How can this chapter help you in your ministry to others?
3. Read verse 5—When did God previously make a statement similar to the one in this verse concerning His Son?
4. How would you describe God's feelings for Jesus?

Record
-Spend three minutes answering the following questions.
1. Write down one question or observation from this chapter. _____
2. List one way this chapter will help you improve your relationship with Jesus. _____
3. List your favorite verse(s) from Matthew 17. _____
4. Write down one action step you will take in your life based on your reading today. *I will* _____

Request
-Devote four minutes to talking with God.

Lesson 34

Relax
  -Spend one minute meditating on a time in your life when God delivered you through a fearful situation.

Read
  -Devote four minutes to reading Psalm 34.

Reflect
  -Take three minutes to think through the following questions.
  1. What are some troubles you are experiencing right now in your life?
  2. What did you read in this Psalm that should comfort you as you face these struggles? (vs. 17-19)
  3. How can this chapter help you minister to others who are facing troubles in their life?
  4. What does it mean to fear God?

Record
  -Spend three minutes answering the following questions.
  1. Write down one question or observation you have over Psalm 34. _____

  2. What is the most important lesson you learned from Psalm 34. _____

  3. Write down the words of verse 14. _____

  4. List one action step you will take in your life based on Psalm 34. *I will* _____

Request
  -Devote four minutes to talking with God.

Lesson 35

Relax
  -Spend one minute meditating on some valuable lessons you have learned from God's Word.

Read
  -Devote four minutes to reading Matthew 18.

Reflect
  -Take three minutes to think through the following questions.
  1. Which verse(s) impacted you the most from this chapter?
  2. What is the most important lesson you learned from Matthew 18?
  3. How can this chapter help you minister to your friends and neighbors?
  4. Is it difficult for you to forgive other people?

Record
  -Spend three minutes answering the following questions.
  1. Write down one question or observation you have over Matthew 18. _____
  _____
  2. List one thing you learned today that will help you in your personal walk with Jesus. _____
  _____
  3. Which servant from verses 21-35 best represents you spiritually? _____
  4. Write out one action step you will take in your life based on today's reading. *I will* _____
  _____

Request
  -Devote four minutes to talking with God.

Lesson 36

Relax
-Spend one minute meditating on two spiritual goals you have set for yourself.

Read
-Devote four minutes to reading Psalm 35.

Reflect
-Take three minutes to think through the following questions.
1. Do you ever feel as if God doesn't hear your prayers?
2. Reflect on a time in your life when a prayer answered no by God turned out to be a blessing.
3. Have you ever felt like David in this chapter?
4. How can this chapter help you in your ministry to this community?

Record
-Spend three minutes answering the following questions.
1. Write down one question or observation you have over Psalm 35. _____
2. How can this Psalm enable you to strengthen your relationship with God? _____
3. How would you describe your relationship with God? _____
4. List one action step you will take in your life based on today's reading. *I will* _____

Request
-Devote four minutes to talking with God.

Lesson 37

Relax
-Spend one minute meditating on the words of Matthew 5:6. *"Blessed are those who hunger and thirst for righteousness, for they shall be satisfied."*

Read
-Devote four minutes to reading Matthew 19.

Reflect
-Take three minutes to think through the following questions.
1. What is the most important lesson you learned from this chapter?
2. How can this chapter help you in your ministry to to your friends and neighbors?
3. What did Jesus mean by His statement in verse 30?

Record
-Spend three minutes answering the following questions.
1. Write down one question or observation you have over this chapter. _____
_____

2. List one way this chapter will strengthen your relationship with Jesus. _____
_____

3. How can this chapter help you improve your relationships with your family? _____
_____

4. List one action step you will take in your life based on today's reading. *I will* _____
_____

Request
-Devote four minutes to talking with God.

Lesson 38

Relax
- Spend one meditating on the words of Psalm 37:4. *"Delight yourself in the Lord and He will give you the desires of your heart."*

Read
- Devote four minutes to reading Psalm 37.

Reflect
- Take three minutes to think through the following questions.
    1. What is the most important lesson you learned from Psalm 37?
    2. What are some things we should delight in spiritually?
    3. What did you learn today that will help you minister to others?
    4. Read verse 23 again. Are the steps of your life established by the Lord?

Record
- Spend three minutes answering the following questions.
    1. Write down one question or observation you have over Psalm 37. _____
    2. List one way this chapter will help you strengthen your relationship with God. _____
    3. Write out one action step you will take in your life based on today's reading. *I will* _____

Request
- Devote four minutes to talking with God.

Lesson 39

Relax
-Spend one minute meditating on how Jesus has shown you compassion.

Read
-Devote four minutes to reading Matthew 20.

Reflect
-Take three minutes to think through the following questions.
1. What is the most important lesson you learned from this chapter?
2. Read verse 34—Are you compassionate towards others as Christ was in His ministry?
3. How can this chapter help you in your ministry to others?
4. How does Jesus measure greatness?

Record
-Spend three minutes answering the following questions.
1. Write down one question or observation you have over Matthew 20. _____

2. List one way this chapter will help you strengthen your relationship with Jesus. _____

3. What are some things you are doing to serve others? _____

4. Write down one action step you will take in your life based on today's reading. *I will* _____

Request
-Devote four minutes to talking with God.

Lesson 40

Relax
-Spend one minute meditating on a time in your life when God has helped you.

Read
-Devote four minutes to reading Psalm 38.

Reflect
-Take three minutes to think through the following questions.
1. What was David troubled by in this Psalm?
2. How does sin affect our relationship with God?
3. Have you ever had thoughts like David expressed in verse 18?
4. How can this chapter help you stay away from sin?

Record
-Spend three minutes answering the following questions.
1. Write down one question or observation you have over Psalm 38. _____

2. List one way this chapter will help you in your service to others. _____

3. Write down one thing you learned today that will strengthen your personal walk with God. _____

4. List one action step you will take in your life based on your reading today. *I will* _____

Request
-Devote four minutes to talking with God asking specifically for strength to overcome temptation.

Lesson 41

Relax
    -Spend one minute meditating on a time in your life when you felt close to God.

Read
    -Devote four minutes to reading Matthew 21.

Reflect
    -Take three minutes to think through the following questions.
1. What amazed you the most about Jesus in this chapter?
2. What is the most important lesson you learned from reading Matthew 21?
3. How can this chapter help you in your daily ministry to the lost?
4. How important is faith/belief in our daily prayer life?

Record
    -Spend three minutes answering the following questions.
1. Write down one question or observation you have from Matthew 21. _____
   _____
2. List one way this chapter will strengthen your personal relationship with Jesus. _____
   _____
3. Write down one action step you will take in your life based on today's reading. *I will* _____
   _____

Request
    -Devote four minutes to talking with God.

Lesson 42

Relax
 -Spend one minute meditating on being able to spend eternity in a place where there is no pain, death or tears.

Read
 -Devote four minutes to reading Psalm 39.

Reflect
 -Take three minutes to think through the following questions.
 1. What will matter most to you when your life is done?
 2. How would your priorities change if you found out you had only one year to live? One month? One week?
 3. How do your current priorities stack up against your answers to question 2?
 4. How can this chapter help you in your personal ministry?

Record
 -Spend three minutes answering the following questions.
 1. Write down one question or observation you have over Psalm 39. _____

 2. List one way this chapter will help you strengthen your relationship with God. _____

 3. Write out one action step you will take in your life based on today's reading. *I will* _____

Request
 -Devote four minutes to talking with God.

Lesson 43

Relax

-Spend one minute meditating on these words from 1st Peter. *"Cast all your anxiety on Him (Jesus) because He cares for you."*

Read

-Devote four minutes to reading Matthew 22.

Reflect

-Take three minutes to think through the following questions.
1. What is the most important lesson you learned from Matthew 22?
2. How important was love to Jesus?
3. How can you develop a stronger love for God and your neighbors?

Record

-Spend three minutes to answer the following questions.
1. Write down one question or observation you have over Matthew 22. _____

2. List one way this chapter will help you improve your relationship with Jesus. _____

3. What did you learn from Matthew 22 that will help you in your service to others? _____

4. Write out one action step you will take in your life based on today's reading. *I will* _____

Request

-Devote four minutes to talking with God.

Lesson 44

Relax
-Spend one minute meditating on the blessing of salvation God has given you and the world through Jesus Christ.

Read
-Devote four minutes to reading Psalm 40.

Reflect
-Take three minutes to think through the following questions.
1. What is the most important lesson you learned from Psalm 40?
2. Which verse(s) in this Psalm describes thoughts you have experienced in your spiritual life?
3. What are some things you desire to do for God?

Record
-Spend three minutes answering the following questions.
1. Write down one question or observation you have over Psalm 40. _____

2. List one way this Psalm will strengthen your personal relationship with God. _____

3. How can verses 9 and 10 of this Psalm help you in your ministry to your relatives, friends and neighbors who are lost? _____

4. Write out one action step you will take in your life based on your reading today. *I will* _____

Request
-Devote four minutes to talking with God.

Lesson 45

Relax
-Spend one minute in a quiet place talking with God about your concerns and goals.

Read
-Devote four minutes to reading Matthew 23.

Reflect
-Take three minutes to think through the following questions.
1. What is the most important lesson you learned from Matthew 23?
2. How would you describe the scribes and Pharisees?
3. What did you learn from this chapter that will help you in your personal ministry to others?

Record
-Spend three minutes answering the following questions.
1. Write down one question or observation you have from Matthew 23. _____

2. List one way this chapter will strengthen your personal relationship with Jesus. _____

3. What is one new thing you learned about Jesus from this chapter? _____

4. Write out one action step you will take in your life based on your reading today. *I will* _____

Request
-Devote four minutes to talking with God.

Lesson 46

Relax
-Spend one minute writing a short note to God expressing your thanks to Him.

Read
-Devote four minutes to reading Psalm 41.

Reflect
-Take three minutes to think through the following questions.
1. Besides Jesus, who is your best friend? Why?
2. What are some characteristics of a good friend?
3. Have you ever been betrayed by a friend?
4. Would you describe yourself as a good friend?
5. Who did the Psalmist turn to when everyone had seemingly turned their back on him?

Record
-Spend three minutes answering the following questions.
1. Write down one question or observation you have from Psalm 41. _____

2. How can this chapter help you strengthen your relationship with God? _____

3. List one thing you learned today that will help you become more effective in your ministry. _____

4. Write down one action step you will take in your life based on Psalm 41. *I will* _____

Request
-Devote four minutes to talking with God.

Lesson 47

Relax
 -Spend one minute meditating on this thought from Titus 1:3. *God will never lie to you!*

Read
 -Devote four minutes to reading Matthew 24.

Reflect
 -Take three minutes to think through the following questions.
 1. What is the greatest promise God has made humanity?
 2. What are some false teachings in religion today concerning the second coming of Christ?
 3. What did you learn from this chapter that will help you in your ministry?

Record
 -Spend three minutes answering the following questions.
 1. Write down one question or observation you have over Matthew 24. _____

 2. What is the most important lesson you learned from reading this chapter? _____

 3. List one way this chapter will strengthen your relationship with Jesus. _____

 4. Write out one action step you will take in your life based on Matthew 24. *I will* _____

Request
 -Devote four minutes to talking with God.

Lesson 48

Relax
-Spend one minute meditating on the words of Psalm 42:1. *"As the deer pants for the water brook, so my soul pants for you O God."*

Read
-Devote four minutes to reading Psalm 42 and Psalm 43.

Reflect
-Take three minutes to think through the following questions.
1. How would you describe your desire to pursue God?
2. What is the most important lesson you learned from these two Psalms?
3. How important is your personal relationship with God when compared to your other relationships?
4. How can this chapter help you become a more effective servant for God?

Record
-Spend three minutes answering the following questions.
1. Write down one question or observation you have over Psalm 42 and 43. _____

2. List one way this chapter will help strengthen your relationship with God. _____

3. Write out one action step you will take in your life based on today's reading. *I will* _____

Request
-Devote four minutes to talking with God.

Lesson 49

Relax
-Spend one minute meditating on how great it feels to be able to help someone.

Read
-Devote four minutes to reading Matthew 25.

Reflect
-Take three minutes to think through the following questions.
1. What is the most important lesson you learned from Matthew 25?
2. Which parable from this chapter impacts you the most? Why?
3. What is something new you learned about the coming judgment day from Matthew 25?
4. How can this chapter help you in your outreach to those who are lost?

Record
-Spend three minutes answering the following questions.
1. Write down one question or observation you have from Matthew 25. _____

2. How can this chapter strengthen your personal commitment to God? _____

3. Write down one action step you will take in your life based on today's reading. *I will* _____

Request
-Devote four minutes to talking with God.

Lesson 50

Relax
-Spend one minute meditating on the words of James 1:12. *"Blessed is a man who perseveres under trial; for once he has been approved, he will receive the crown of life which the Lord has promised to those who love Him."*

Read
-Devote four minutes to reading Psalm 44.

Reflect
-Take three minutes to think through the following questions.
1. Who do you trust in your life? Do you trust God?
2. What are some things you have learned from your elders?
3. What was David asking God for in this Psalm?
4. Who do you turn to when you are facing a trial in your life?

Record
-Spend three minutes answering the following questions.
1. Write down one question or observation you have from this Psalm. _____

2. List one way this Psalm will strengthen your relationship with God. _____

3. Write down one action step you will take in your life based on today's reading. *I will* _____

Request
-Devote four minutes to talking with God asking specifically for strength to overcome the trials you are currently facing.

Lesson 51

Relax
    -Spend one minute in a quiet place praying specifically for your family.

Read
    -Devote four minutes to reading Matthew 26.

Reflect
    -Take three minutes to think through the following questions.
1. What is the most important lesson you learned from Matthew 26?
2. How would you describe Jesus based on what you read about Him in verses 36-46?
3. How difficult is it for you to give up control of your life to God?
4. What did you learn about submitting your will to God from Jesus?

Record
    -Spend three minutes answering the following questions.
1. Write down one question or observation you have from Matthew 26. _____
2. How can this chapter help you strengthen your relationship with God? _____
3. Write down one action step you will take in your life based on today's reading. *I will* _____

Request
    -Devote four minutes to talking with God.

Lesson 52

Relax
 -Spend one minute following the words of Psalm 46:10. *"Be still and know that I am God."*

Read
 -Devote four minutes to reading Psalm 46 and Psalm 47.

Reflect
 -Take three minutes to think through the following questions.
   1. What is the most important lesson you learned from these two Psalms?
   2. What are some things people fear in their lives?
   3. What did you learn about fear from Psalm 46:2?
   4. Why should we as Christians not get bogged down with fear?

Record
 -Spend three minutes answering the following questions.
   1. Write down one question or observation you have from Psalm 46 and Psalm 47. _____
   2. List one way these chapters have improved your understanding of who God is. _____
   3. Write down two reasons why we should want to praise God. _____
   4. Write down one action step you will take in your life based on today's reading. *I will* _____

Request
 -Devote four minutes to talking with God.

Lesson 53

Relax
    -Spend one minute meditating on the immeasurable love Jesus has for you as shown in His death for you on the cross.

Read
    -Devote four minutes to reading Matthew 27.

Reflect
    -Take three minutes to think through the following questions.
1. What is the most important lesson you learned from Matthew 27?
2. How would you describe Jesus based on what you see about Him from this chapter?
3. What do you think was the most difficult aspect of the cross for Jesus?

Record
    -Spend three minutes answering the following questions.
1. Write down one question or observation you have over Matthew 27. _____

2. List one way this chapter will strengthen your relationship with Jesus. _____

3. What did you learn from Jesus in this chapter that will help you become a better servant to others? ___

4. Write down one action step you will take in your life based on your reading today. *I will* _____

Request
    -Devote four minutes to talking with God.

Lesson 54

Relax
   -Spend one minute meditating on the words of Isaiah 40:8. *"The grass withers, the flower fades, but the word of our God stands forever."*

Read
   -Devote four minutes to reading Psalm 48.

Reflect
   -Take three minutes to think through the following questions.
   1. What is the most important lesson you learned from Psalm 48?
   2. If you had to describe God in one word, what word would you choose? Why?
   3. How would you describe the difference that God has made in your life?

Record
   -Spend three minutes answering the following question.
   1. Write down one question or observation you have from Psalm 48. _____
   2. List one way this chapter has improved your understanding of God. _____
   3. What amazes you the most about God? _____
   4. Write out one action step you will take in your life based on your reading today. *I will* _____

Request
   -Devote four minutes to talking with God.

Lesson 55

Relax
    -Spend one minute meditating on the victory Jesus brings us over sin and death through His death and resurrection.

Read
    -Devote four minutes to reading Matthew 28.

Reflect
    -Take three minutes to think through the following questions.
1. What is the most important lesson you learned from this chapter?
2. Why did the elders pay the soldiers to lie about the events that happened at Jesus' tomb?
3. Why is Matthew 28:18-20 generally referred to as the Great Commission?

Record
    -Spend three minutes answering the following questions.
1. Write down one question or observation you have over Matthew 28. _____

2. List one way this chapter will strengthen your relationship with Jesus. _____

3. How can Matthew 28 help you in your ministry to others? _____

4. Write out one action step you will take in your life based on Matthew 28. *I will* _____

Request
    -Devote four minutes to talking with God.

Lesson 56

Relax
-Spend one minute praying to God asking Him for strength to fully trust in Him.

Read
-Devote four minutes to reading Psalm 49.

Reflect
-Take three minutes to think through the following questions.
1. What is the most important lesson you learned from this chapter?
2. What are some things today that people trust in besides God?
3. How can we learn to trust in God?
4. What are some things God has done for you that should cause you to trust in Him?

Record
-Spend three minutes answering the following questions.
1. Write down one question or observation you have over Psalm 49. _____
2. List one way this chapter will improve your personal relationship with God. _____
3. Describe the importance of understanding who God is. _____
4. Write out one action step you will take in your life based on today's reading. *I will* _____

Request
-Devote four minutes to talking with God.

Lesson 57

Relax
    -Spend one minute meditating on the fact that Jesus will someday return to earth to take the faithful home to be with Him in heaven for eternity.

Read
    -Devote four minutes to reading Acts 1.

Reflect
    -Take three minutes to think through the following questions.
1. What is the most significant lesson you learned from Acts 1?
2. How do you think the apostles felt when watching Jesus ascend into heaven before their very eyes?
3. How do you feel knowing Jesus will return to judge all humanity?
4. Why did Jesus stay on earth for forty days after His resurrection?

Record
    -Spend three minutes answering the following questions.
1. Write down one question or observation you have over Acts 1. _____
2. List one way this chapter will help you strengthen your relationship with Jesus. _____
3. Write out one action step you will take in your life based on today's reading. *I will* _____

Request
    -Devote four minutes to talking with God.

Lesson 58

Relax
-Spend one minute meditating on the power of God displayed in Genesis 1:1. *"In the beginning, God created the heavens and the earth."*

Read
-Devote four minutes to reading Psalm 50.

Reflect
-Take three minutes to think through the following questions.
1. What is the most important lesson you learned from Psalm 50?
2. How does God feel about those who are full of wickedness?
3. How would you describe the power of God?

Record
-Spend three minutes answering the following questions.
1. Write down one question or observation you have over Psalm 50. _____

2. List one quality you learned about God that will improve your relationship with Him. _____

3. Which sin from verses 17-20 do you struggle the most with in your life? _____

4. Write down one action step you will take in your life based on Psalm 50. *I will* _____

Request
-Devote four minutes to talking with God.

Lesson 59

Relax
  -Spend one minute in prayer to God asking Him for courage to stand up for Him each day.

Read
  -Devote four minutes to reading Acts 2.

Reflect
  -Take three minutes to think through the following questions.
  1. What is the most important lesson you learned from this chapter?
  2. How would you describe the sermon Peter preached in this chapter?
  3. What do we learn about Jesus from verse 36?
  4. Why was it necessary for the apostles to speak different languages?

Record
  -Spend three minutes answering the following questions.
  1. Write down one question or observation you have from Acts 2. _____

  2. List one way this chapter will strengthen your relationship with Jesus. _____

  3. What is the purpose of baptism as seen in verse 38?
  _____

  4. Write down one action step you will take in your life based on your reading today. *I will* _____
  _____

Request
  -Devote four minutes to talking with God.

Lesson 60

Relax
    -Spend one minute meditating on the great compassion God shows us through forgiving us of our sins.

Read
    -Devote four minutes to reading Psalm 51.

Reflect
    -Take three minutes to think through the following questions.
1. What is the most significant lesson you learned from Psalm 51?
2. Have you ever tried to earn God's forgiveness and acceptance?
3. What do you think a broken spirit and a contrite heart looks like?

Record
    -Spend three minutes answering the following questions.
1. Write down one question or observation you have over Psalm 51. _____
2. List one way this chapter will strengthen your relationship with God. _____
3. How can this chapter help you reach out to those who are trapped in sin? _____
4. Write down one action step you will take in your life based on your reading today. *I will* _____

Request
    -Devote four minutes to talking with God in prayer.

Lesson 61

Relax
 -Spend one minute in a quiet place shifting your thoughts towards God.

Read
 -Devote four minutes to reading Acts 3.

Reflect
 -Take three minutes to think through the following questions.
 1. What is the most important lesson you learned from Acts 3?
 2. How were John and Peter able to heal the lame man in this chapter?
 3. What message did Peter preach in his sermon in this chapter?
 4. What does it mean to repent? (Compare your thoughts with verse 19.)

Record
 -Spend three minutes answering the following questions.
 1. Write down one question or observation you have over Acts 3. _____

 2. List one way this chapter will strengthen your relationship with Jesus. _____

 3. Write out one action step you will take in your life based on your reading today. *I will* _____

Request
 -Devote four minutes to talking with God.

Lesson 62

Relax
 -Spend one minute meditating on Psalm 54:4. *"God is my helper; the Lord is the sustainer of my soul."*

Read
 -Devote four minutes to reading Psalm 52 – Psalm 54.

Reflect
 -Take three minutes to think through the following questions.
 1. What is the most important lesson you learned from these Psalms?
 2. How do you feel knowing God wants to help you?
 3. How would you describe the love of God? (Compare your thoughts with Psalm 52:1)
 4. Why do you believe in God?

Record
 -Spend three minutes answering the following questions.
 1. Write down one question or observation you have from these Psalms. _____
 2. List one way these Psalms will improve your personal walk with God. _____
 3. Write down the name of one of your friends who needs to know Jesus and begin praying for them.
 4. Write out one action step you will take in your life based on your reading today. *I will* _____

Request
 -Devote four minutes to talking with God.

Lesson 63

Relax
-Spend one minute in a quiet place talking with God about concerns you have in your life.

Read
-Devote four minutes to reading Acts 4.

Reflect
-Take three minutes to think through the following questions.
1. What is the most important lesson you learned from reading Acts 4?
2. Why did the early church experience such rapid growth?
3. Where did Peter and John get the courage to share the message of Jesus?

Record
-Spend three minutes answering the following questions.
1. Write down one question or observation you have over Acts 4. _____
   _____

2. List one way this chapter will strengthen your relationship with Jesus. _____
   _____

3. What is one lesson you learned about serving others from this chapter? _____
   _____

4. Write down one action step you will take in your life based on today's reading. *I will* _____
   _____

Request
-Devote four minutes to talking with God.

Lesson 64

Relax

-Spend one minute meditating on the words found in Psalm 55:22. *"Cast your burden upon the Lord and He will sustain you."*

Read

-Devote four minutes to reading Psalm 55.

Reflect

-Take three minutes to think through the following questions.
1. What is the most powerful lesson you learned from reading Psalm 55?
2. Has there ever been a time in your life when you doubted the words of Psalm 55:22?
3. What are the three biggest concerns in your life right now?
3. How did David handle the doubts and concerns he faced in his life?

Record

-Spend three minutes answering the following questions.
1. Write down one question or observation you have over Psalm 55. _____

2. List one way this chapter will improve your relationship with God. _____

3. Write out one action step you will take in your life based on your reading today. *I will* _____

Request

-Devote four minutes to talking with God.

Lesson 65

Relax
-Spend one minute in a quiet place writing down one spiritual goal you want to achieve in the upcoming year.

Read
-Devote four minutes to reading Acts 5.

Reflect
-Take three minutes to think through the following questions.
1. What is the most important lesson you learned from Acts 5?
2. Why were Ananias and Sapphira struck down by God?
3. What impact did their deaths have on the growth of the early church?
4. Are you living the statement Peter made in verse 29?

Record
-Spend three minutes answering the following questions.
1. Write out one question or observation you have over Acts 5. _____

2. List one way this chapter will improve your relationship with God. _____

3. In what places did the apostles preach the gospel. _____

4. Write out one action step you will take in your life based on your reading today. *I will* _____

Request
-Devote four minutes to talking with God.

Lesson 66

Relax
-Spend one minute thanking God for the blessings He has given you.

Read
-Devote four minutes to reading Psalm 56 and Psalm 57.

Reflect
-Take three minutes to think through the following questions.
1. What is the most important lesson you learned from these Psalms?
2. What are some things David was thankful for in these Psalms?
3. Are people sometimes guilty of taking God for granted? Explain.

Record
-Spend three minutes answering the following questions.
1. Write down one question or observation you have over these Psalms. _____

2. List one thing you learned today that will strengthen your commitment to God. _____

3. How can you bring God praise for all the things He has done for you? _____

4. Write out one action step you will take in your life based on your reading today. *I will* _____

Request
-Devote four minutes to talking with God.

Lesson 67

Relax
-Spend one minute meditating on the words of James 4:8. *"Come near to God, and He will come near to you."*

Read
-Devote four minutes to reading Acts 6.

Reflect
-Take three minutes to think through the following questions.
1. What is the most important lesson you learned from Acts 6?
2. How would you describe the problem that arose in this chapter?
3. What steps did the apostles' take to solve this problem?
4. What happened as a result of the problem being addressed and solved?

Record
-Spend three minutes answering the following questions.
1. Write out one question or observation you have over Acts 6. _____
2. List one way this chapter will help you in your Christian walk. _____ _____
3. List the two things the apostles spent their time doing. _____, _____
4. Write out one action step you will take in your life based on today's reading. *I will* _____ _____

Request
-Devote four minutes to talking with God.

Lesson 68

Relax
-Spend one minute meditating on this thought. *I was created to become like Christ—Colossians 1:28.*

Read
-Devote four minutes to reading Psalm 59.

Reflect
-Take three minutes to think through the following questions.
1. What is one important lesson you learned from Psalm 59?
2. Read Matthew 5:44-48: How does Jesus want us to treat our enemies?
3. Is it difficult for you to fulfill this command? Why?
4. Where do you turn in your life for strength?
5. In what ways has God shown you His lovingkindness?

Record
-Spend three minutes answering the following questions.
1. Write out one question or observation you have over Psalm 59. _____

2. List one way this chapter will strengthen your relationship with God. _____

3. Write out one action step you will take in your life based on today's reading. *I will* _____

Request
-Devote four minutes to talking with God.

Lesson 69

Relax
 -Spend one minute meditating on the words found in Numbers 6:25. *"The Lord makes His face shine on you, and be gracious to you."*

Read
 -Devote four minutes to reading Acts 7.

Reflect
 -Take three minutes to think through the following questions.
 1. What is one important lesson you learned today?
 2. How would you describe Stephen's message?
 3. What is one new thing you learned about the history of the Israelites from this chapter?
 4. Read verse 60: How was Stephen able to make this statement about his killers?
 5. Who else made a similar statement to Stephen's concerning the people who put Him to death?

Record
 -Spend three minutes answering the following questions.
 1. Write down one question or observation you have over Acts 7. _____

 2. List one way this chapter will improve your commitment to God. _____

 3. Write out one action step you will take in your life based on today's reading. *I will* _____

Request
 -Devote four minutes to talking with God.

Lesson 70

Relax
-Spend one minute meditating on this thought from Psalm 149. *"The Lord takes pleasure in you."*

Read
-Devote four minutes to reading Psalm 60 and Psalm 61.

Reflect
-Take three minutes to think through the following questions.
1. What is the most important lesson you learned from these two Psalms?
2. What do you think David had experienced prior to writing Psalm 60?
3. How does David describe God in Psalm 61?
4. Do you try to handle the struggles in your life without God's help?
5. Why is it sometimes difficult to turn to God for help in our lives?

Record
-Spend three minutes answering the following questions.
1. Write out one question or observation you have over Psalm 60 and Psalm 61. _____

2. List one way this chapter will strengthen your relationship with God. _____

3. Write out one action step you will take in your life based on today's reading. *I will* _____

Request
-Devote four minutes to talking with God.

Lesson 71

Relax
    -Spend one minute meditating on this thought. *"You were made for a mission."*

Read
    -Devote four minutes to reading Acts 8.

Reflect
    -Take three minutes to think through the following questions.
1. What is the most important lesson you learned from this chapter?
2. How did the persecution against the church turn out to be a blessing?
3. What are some persecutions/sufferings you face in your life as a Christian?

Record
    -Spend three minutes answering the following questions.
1. Write down one question or observation you have over Acts 8. _____

2. List one way this chapter will strengthen your relationship with Jesus. _____

3. What did you learn from Philip that will help you in your service to others? _____

4. Write out one action step you will take in your life based on today's reading. *I will* _____

Request
    -Devote four minutes to talking with God.

Lesson 72

Relax
-Spend one minute meditating on the positive changes Jesus has brought into your life.

Read
-Devote four minutes to reading Acts 9.

Reflect
-Take three minutes to think through the following questions.
1. What is the most important lesson you learned from this chapter?
2. How would you describe the changes in Saul's life after becoming a Christian?
3. In what ways did your life change after you became a Christian?
4. What were some obstacles Saul faced in his ministry?

Record
-Spend three minutes answering the following questions.
1. Write down one question or observation you have over Acts 9. _____
2. List one way this chapter will strengthen your walk with Christ. _____
3. Write out one action step you will take in your life based on today's reading. *I will* _____

Request
-Devote four minutes to talking with God.

Lesson 73

Relax
  -Spend one minute meditating on the greatness of God.

Read
  -Devote four minutes to reading Psalm 62 and Psalm 63.

Reflect
  -Take three minutes to think through the following questions.
  1. What is the most important lesson you learned from these two Psalms?
  2. Read Psalm 62:1—How can someone wait in silence for God?
  3. Why should someone put their trust in God?
  4. What are some things you have put your trust in besides God?

Record
  -Spend three minutes answering the following questions.
  1. Write down one question or observation you have over Psalm 62 and 63. _____
  2. List one way these Psalms will improve your daily relationship with God. _____
  3. In one sentence, describe what God means to you in your life. _____
  4. List one action step you will take in your life based on today's reading. *I will* _____

Request
  -Devote four minutes to talking with God.

Lesson 74

Relax
-Spend one minute in a quiet place talking with God about your day.

Read
-Devote four minutes to reading Acts 10.

Reflect
-Take three minutes to think through the following questions.
1. What is the most significant lesson you learned from Acts 10?
2. What role did prayer play in the conversion of Cornelius and his family?
3. How would you describe your prayer life?
4. What was the significance of this family being given the opportunity to respond to the gospel? (vs. 45)

Record
-Spend three minutes answering the following questions.
1. Write out one question or observation you have over Acts 10. _____

2. List one way this chapter will strengthen your walk with Jesus. _____

3. The message of the gospel is for _____.
4. List one action step you will take in your life based on today's reading. *I will* _____

Request
-Devote four minutes to talking with God.

Lesson 75

Relax
 -Spend one minute meditating on the words of James 5:11b.
 "The Lord is full of compassion and mercy."

Read
 -Devote four minutes to reading Acts 11.

Reflect
 -Take three minutes to think through the following questions.
 1. What is the most important lesson you learned from this chapter?
 2. How would you describe the problem Peter faced in this chapter?
 3. How did Peter handle this problem?
 4. What can we learn from this chapter about dealing with the problems we face in our lives?

Record
 -Spend three minutes answering the following questions.
 1. Write down one question or observation you have over Acts 11. _____

 2. List one way this chapter will improve your commitment to God. _____

 3. What can we learn about serving others from Barnabas? _____
 4. Write out one action step you will take in your life based on today's reading. *I will* _____

Request
 -Devote four minutes to talking with God.

Lesson 76

Relax
  -Spend one minute in a quiet place meditating on the things that matter the most in your life.

Read
  -Devote four minutes to reading Psalm 64 and Psalm 65.

Reflect
  -Take three minutes to think through the following questions.
  1. What is the most important lesson you learned from these Psalms?
  2. In what ways has God taking care of you?
  3. How has God taken care of your spiritual needs?

Record
  -Spend three minutes answering the following questions.
  1. Write down one question or observation you have from these two Psalms. _____

  2. List one way this chapter will strengthen your personal relationship with God. _____

  3. Write out the words of the verse that impacted you the most from these two Psalms. _____

  4. Write out one action step you will take in your life based on today's reading. *I will* _____

Request
  -Devote four minutes to talking with God in prayer.

Lesson 77

Relax
    -Spend one minute in a quiet place expressing your feelings and emotions to God, your Father. *Tell Him how you feel and what you feel!*

Read
    -Devote four minutes to reading Acts 12.

Reflect
    -Take three minutes to think through the following questions.
1. What is the most interesting thing you read in this chapter?
2. What do you think Peter was thinking while sitting in prison?
3. What role did prayer play in the release of Peter?
4. When you are faced with a difficult situation, do you rely on the power of prayer to get through it?

Record
    -Spend three minutes answering the following questions.
1. Write down one question or observation you have over Acts 12. _____

2. List one way this chapter will deepen your commitment to Jesus. _____

3. Write out one action step you will take in your life based on today's reading. *I will* _____

Request
    -Devote four minutes to talking with God.

Lesson 78

Relax
-Spend one minute writing down two spiritual goals you have reached in your life.
1.
2.

Read
-Devote four minutes to reading Acts 13.

Reflect
-Take three minutes to think through the following questions.
1. What is the most important lesson you learned from this chapter?
2. What two things did Paul and Barnabas do before leaving on their mission trip?
3. Should we as Christians fast today?
4. What did you learn in this chapter that will make you a more effective servant?

Record
-Spend three minutes answering the following questions.
1. Write down one question or observation you have over Acts 13. _____

2. List one way this chapter will help you in your daily walk with Jesus. _____

3. List one action step you will take in your life based on today's reading. *I will* _____

Request
-Devote four minutes to talking with God.

Lesson 79

Relax
-Spend one minute meditating on the awesome works of God. *"How awesome are your works!" Psalm 66:3*

Read
-Devote four minutes to reading Psalm 66 and Psalm 67.

Reflect
-Take three minutes to think through the following questions.
1. What is the most significant lesson you learned from these Psalms?
2. What are some of the most impressive things God has created?
3. Read Psalm 66:10—How does God refine us in our lives today?
4. How should Psalm 67:2 impact your ministry?

Record
-Spend three minutes answering the following questions.
1. Write down one question or observation you have over these Psalms. _____

2. List one way this chapter will improve your relationship with God. _____

3. What are you doing in your daily life to bring praise to God? _____
4. Write out one action step you will take in your life based on today's reading. *I will* _____

Request
-Devote four minutes to talking with God.

Lesson 80

Relax
-Spend one minute meditating on the words of Philippians 4:6. *"Be anxious in nothing, but in everything by prayer and supplication make your requests known to God."*

Read
-Devote four minutes to reading Acts 14.

Reflect
-Take three minutes to think through the following questions.
1. What is the most important lesson you learned from this chapter?
2. If you had the opportunity to go on a missionary effort with Paul of the NT, would you go?
3. How would you describe the experiences of Paul and Barnabas in this chapter?

Record
-Spend three minutes answering the following questions.
1. Write down one question or observation you have over this chapter. _____

2. List one way this chapter will strengthen your commitment to Jesus. _____

3. What quality do you admire the most about Paul in this chapter? _____
4. Write out one action step you will take in your life based on today's reading. *I will* _____

Request
-Devote four minutes to talking with God in prayer.

Lesson 81

Relax
-Spend one minute in a quiet place thanking God for the blessing of salvation.

Read
-Devote four minutes to reading Acts 15.

Reflect
-Take three minutes to think through the following questions.
1. What is the most interesting thing you learned from this chapter?
2. What is the difference between following tradition and following God's commands?
3. How can following traditions cause problems?
4. Who do you think was right in the argument between Paul and Barnabas over John Mark? Why?

Record
-Spend three minutes answering the following questions.
1. Write out one question or observation you have over this chapter. _____

2. In what ways, do we all need to be like Barnabas? _____

3. List one way this chapter will strengthen your daily walk with Jesus. _____

4. List one action step you will take in your life based on today's reading. *I will* _____

Request
-Devote four minutes to talking with God.

Lesson 82

Relax
-Spend one minute meditating on the words of Psalm 68:19. *"Blessed be the Lord, who daily bears our burden."*

Read
-Devote four minutes to reading Psalm 68.

Reflect
-Take three minutes to think through the following questions.
1. What is the most interesting thing you learned from this chapter?
2. Read verse 20: In what ways has God provided deliverance for you?
3. Why did David describe God as being awesome?

Record
-Spend three minutes answering the following questions.
1. Write down one question or observation you have over Psalm 68. _____
2. List one way this chapter will strengthen your relationship with God. _____
3. Write out the words of the verse from this chapter that impacted you the most today. _____
4. Write down one action step you will take in your life based on today's reading. *I will* _____

Request
-Devote four minutes to talking with God.

Lesson 83

Relax
   -Spend one minute meditating on the words of Matthew 5:8.
   "Blessed are the pure in heart, for they shall see God."

Read
   -Devote four minutes to reading Acts 16.

Reflect
   -Take three minutes to think through the following questions.
   1. What is the most exciting thing you read in this chapter?
   2. Have you ever been persecuted because of your faith in Jesus?
   3. On a scale of 1-10, how passionate are you about serving Jesus Christ?

Record
   -Spend three minutes answering the following questions.
   1. Write down one question or observation you have over Acts 16. _____

   2. List one way this chapter will deepen your commitment to Jesus. _____

   3. How will this chapter help you become a more effective Christian? _____

   4. Write down one action step you will take in your life based on today's reading. *I will* _____

Request
   -Devote four minutes to talking with God.

Lesson 84

Relax
-Spend one minute in a quiet place sharing your dreams with God.

Read
-Devote four minutes to reading Acts 17.

Reflect
-Take three minutes to think through the following questions.
   1. What is the most important lesson you learned from this chapter?
   2. Are you influencing people to follow Christ based on your actions?
   3. Read verse 28: How do you feel knowing you are God's child?

Record
-Spend three minutes answering the following questions.
   1. Write out one question or observation you have over this chapter. _____

   2. List one way this chapter will help you in your daily walk with Jesus. _____

   3. What is one thing you can work on to become a better parent or child? _____

   4. Write out one action step you will take in your life based on today's reading. *I will* _____

Request
-Devote four minutes to talking with God.

Lesson 85

Relax
-Spend one minute meditating on the words of 2nd Corinthians 12:9. *"My grace is sufficient for you, my power is made perfect in weakness."*

Read
-Devote four minutes to reading Psalm 69.

Reflect
-Take three minutes to think through the following questions.
1. What is the most important lesson you learned from this chapter?
2. Read verse 6: What was David concerned about in this verse?
3. What kind of reputation do you have with your family and friends?
4. What impact do your actions have on other people?

Record
-Spend three minutes answering the following questions.
1. Write down what you consider to be your greatest strength as a Christian. _____
2. What is one area you feel you need to grow in as a Christian? _____
3. List one way this chapter will strengthen your daily relationship with God. _____
   _____
4. Write out one action step you will take in your life based on today's reading. *I will* _____
   _____

Request
-Devote four minutes to talking with God.

Lesson 86

Relax
-Spend one minute meditating on the power of God's Word. *"For the Word of God is living and active and sharper than any two-edged sword."*

Read
-Devote four minutes to reading Acts 18.

Reflect
-Take three minutes to think through the following questions.
1. What is the most important lesson you learned from Acts 18?
2. How will this chapter help you gain the courage to continue serving God?
3. Read verse 5: How would you describe the impact God's Word had on Paul?
4. How strong is your devotion to God's Word?

Record
-Spend three minutes answering the following questions.
1. What important area of ministry do we see in action in the lives of Priscilla and Aquila? _____
2. List one way this chapter will strengthen your relationship with Jesus. _____
3. List one action step you will take in your life based on today's reading. *I will* _____

Request
-Devote four minutes to talking with God.

Lesson 87

Relax
    -Spend one minute reflecting on how you felt the day you became a Christian.

Read
    -Devote four minutes to reading Acts 19.

Reflect
    -Take three minutes to think through the following questions.
1. What is the most interesting thing you learned from this chapter?
2. What role did miracles have in Paul's ministry?
3. Why were the men at the beginning of this chapter baptized again?

Record
    -Spend three minutes answering the following questions.
1. Write down one question or observation you have over Acts 19. _____
2. List one way this chapter will improve your commitment to Jesus. _____
3. Write down the name of a friend who needs to hear your story about how you became a Christian.
4. List one action step you will take in your life based on today's reading. *I will* _____

Request
    -Devote four minutes to talking with God.

Lesson 88

Relax
 -Spend one minute meditating on the following quote.
 "Your identity is in eternity. Your homeland is in heaven."

Read
 -Devote four minutes to reading Psalm 70 and Psalm 71.

Reflect
 -Take three minutes to think through the following questions.
 1. What is the most important lesson you learned from these Psalms?
 2. What are some things you hope for in your life?
 3. What are some results of placing our complete hope in God?
 4. How difficult is it for you to trust in the Lord?

Record
 -Spend three minutes answering the following questions.
 1. Write down one question or observation you have over these Psalms. _____

 2. List one way this chapter has improved your overall understanding of who God is. _____

 3. Write down one action step you will take in your life based on today's reading. *I will* _____
 _____

Request
 -Devote four minutes to talking with God.

Lesson 89

Relax
-Spend one minute meditating on the words of Romans 8:28. *"And we know that God causes all things to work together for good to those who love God, and are called according to His purpose."*

Read
-Devote four minutes to reading Acts 20.

Reflect
-Take three minutes to think through the following questions.
1. What is the most important lesson you learned from this chapter?
2. How would you describe the purpose of Paul's life?
3. What is the purpose of your life?
4. Are your desires the same as God's desires for you?

Record
-Spend three minutes answering the following questions.
1. Write down one question or observation you have over this chapter. _____

2. What did you learn about Paul from verse 37? _____

3. List one way this chapter will strengthen your relationship with Jesus. _____

4. Write out one action step you will take in your life based on today's reading. *I will* _____

Request
-Devote four minutes to talking with God.

Lesson 90

Relax
-Spend one minute in a quiet place focusing your thoughts on God.

Read
-Devote four minutes to reading Acts 21.

Reflect
-Take three minutes to think through the following questions.
1. What is the most interesting thing you learned from this chapter?
2. Why did God allow Paul to endure suffering and trials while preaching the gospel?
3. What are some trials you are currently facing?
4. Why does God allow us to face trials and struggles as Christians?

Record
-Spend three minutes answering the following questions.
1. Write down one question or observation you have over this chapter. _____

2. List one way this chapter will deepen your commitment to God. _____

3. What is one quality of Paul you need to imitate in your life? _____
4. Write out one action step you will take in your life based on today's reading. *I will* _____

Request
-Devote four minutes to talking with God.

Lesson 91

Relax
-Spend one minute talking with God about your relationship with Him.

Read
-Devote four minutes to reading Psalm 72.

Reflect
-Take three minutes to think through the following questions.
1. What is the most important lesson you learned from this Psalm?
2. How would you describe David's relationship with God?
3. How would you describe your relationship with God?
4. What are some benefits of having a close relationship with God?

Record
-Spend three minutes answering the following questions.
1. Write down one question or observation you have over this Psalm. _____
2. List one way this Psalm will help you in your daily walk with God. _____
3. Write out one action step you will take in your life based on today's reading. *I will* _____

Request
-Devote four minutes to talking with God.

Lesson 92

Relax
-Spend one minute in a quiet place focusing your thoughts and attention towards God.

Read
-Devote four minutes to reading Acts 22.

Reflect
-Take three minutes to think through the following questions.
1. What is the most interesting thing you learned from this chapter?
2. How did Paul defend himself before his accusers?
3. How did the crowd respond to Paul's message?
4. Are people today sometimes like the Romans that Paul addressed in this chapter?

Record
-Spend three minutes answering the following questions.
1. Write down one question or observation you have over Acts 22. _____

2. List one way this chapter will strengthen your personal relationship with Jesus. _____

3. List one talent God has blessed you with that you can use to help the church grow. _____

4. Write out one action step you will take in your life based on today's reading. *I will* _____

Request
-Devote four minutes to talking with God.

Lesson 93

Relax
-Spend one minute meditating on a time in your life when you courageously stood up for Jesus.

Read
-Devote four minutes to reading Acts 23.

Reflect
-Take three minutes to think through the following questions.
1. What is the most important lesson you learned from this chapter?
2. Read verse 1: At this point in your life, can you make this same statement to God that Paul made?
3. What are some things that sometimes prevent you from fulfilling your duty to God?
4. Why was it necessary for Paul to go to Rome?

Record
-Spend three minutes answering the following questions.
1. Write down one question or observation you have over Acts 23. _____
_____

2. List one way this chapter will help you strengthen your relationship with Jesus. _____
_____

3. What was the source of Paul's courage? _____
_____

4. Write out one action step you will take in your life based on today's reading. *I will* _____
_____

Request
-Devote four minutes to talking with God.

Lesson 94

Relax
-Spend one minute meditating on the words of James 4:7. *"Submit therefore to God. Resist the devil and he will flee from you."*

Read
-Devote four minutes to reading Psalm 73.

Reflect
-Take three minutes to think through the following questions.
1. What temptation was the psalmist facing in this chapter?
2. What are two temptations you are currently facing in your life?
3. How can we resist the devil?
4. What is your action plan to overcome the temptations you are facing in your life?

Record
-Spend three minutes answering the following questions.
1. Write down one question or observation you have over Psalm 73. _____

2. List one way God's Word has helped you overcome past temptations. _____

3. Write out one action step you will take in your life based on today's reading. *I will* _____

Request
-Devote four minutes to talking with God.

Lesson 95

Relax
-Spend one minute meditating on God's definition of love found in 1st Corinthians 13:4-8.

Read
-Devote four minutes to reading Acts 24.

Reflect
-Take three minutes to think through the following questions.
1. What is the most interesting lesson you learned from this chapter?
2. Read verse 21: Why did Paul say he was put on trial?
3. How would you describe Paul's love for God?
4. How can we tell whether or not someone is in love with God?
5. Based on your actions, do you love God?

Record
-Spend three minutes answering the following questions.
1. Write down one question or observation you have over this chapter. _____

2. List one thing you learned from Paul in Acts 24 tha will help you deepen your commitment to Jesus. ___

3. Write out one action step you will take in your life based on today's reading. *I will* _____

Request
-Devote four minutes to talking with God.

Lesson 96

Relax
-Spend one minute in a quiet place talking with God about your day.

Read
-Devote four minutes to reading Acts 25.

Reflect
-Take three minutes to think through the following questions.
1. What are some events in life that can sometimes rock our faith?
2. How was Paul able to maintain his faith through all the trials and struggles he faced?
3. What is the greatest struggle you have endured as a Christian?

Record
-Spend three minutes answering the following questions.
1. Write down one question or observation you have over Acts 25. _____

2. List one way this chapter will strengthen your relationship with Jesus. _____

4. Write out one action step you will take in your life based on today's reading. *I will* _____

Request
-Devote four minutes to talking with God.

Lesson 97

Relax
-Spend one minute meditating on the words of Ephesians 3:20. *The dreams of the biggest dreamer are way too small in the eyes of God.*

Read
-Devote four minutes to reading Psalm 74.

Reflect
-Take three minutes to think through the following questions.
1. How would you describe the power of God?
2. What do we learn about the power of God from this Psalm?
3. Why do often fail to let our all-powerful Father have complete control of our lives?
4. Reflect on the following statement. *You will never find God until you forget about self.*

Record
-Spend three minutes answering the following questions.
1. Write down one question or observation you have over Psalm 74. _____

2. List one way this chapter will strengthen your relationship with God. _____

3. List one action step you will take in your life based on today's reading. *I will* _____

Request
-Devote four minutes to talking with God.

Lesson 98

Relax
-Spend one minute reflecting on the opportunity for repentance God continually offers us as Christians.

Read
-Devote four minutes to reading Acts 26.

Reflect
-Take three minutes to think through the following questions.
1. What is the most interesting lesson you learned from this chapter?
2. What do you think prevented Festus from becoming a Christian?
3. What are some things that prevent people from obeying the gospel today?
4. How will this chapter help you in your ministry to the lost?

Record
-Spend three minutes answering the following questions.
1. Write down one question or observation you have over this chapter. _____

2. List one thing you learned today that will help you in your daily walk with Jesus. _____

3. Write out one action step you will take in your life based on today's reading. *I will* _____

Request
-Devote four minutes to talking with God.

Lesson 99

Relax
   -Spend one minute in a quiet place reading Paul's words to Timothy in 2nd Timothy 4:6-8.

Read
   -Devote four minutes to reading Acts 27.

Reflect
   -Take three minutes to think through the following questions.
   1. What is the most interesting lesson you learned from this chapter?
   2. How would you describe Paul's journey to Rome?
   3. In what ways is our spiritual life like a journey?
   4. What did you learn from Paul that will help you handle the rough aspects of your Christian journey?

Record
   -Spend three minutes answering the following questions.
   1. Write down one question or observation you have over Acts 27. _____
   2. Write the name of the person who has influenced you the most to live for Christ. _____
   3. List one way this chapter will help you stay motivated to live for Jesus. _____
   4. Write out one action step you will take in your life based on today's reading. *I will* _____

Request
   -Devote four minutes to talking with God.

Lesson 100

Relax
-Spend one minute reflecting on how your life has been blessed by the time you have spent with God these past 100 days.

Record
-Devote four minutes to reading Acts 28.

Reflect
-Take three minutes to think through the following questions.
1. What is the most interesting lesson you learned from this chapter?
2. How did Paul spend his time while in Rome?
3. How would you describe the impact that Paul had on the people he met in Rome?

Record
-Spend three minutes answering the following questions.
1. Write down one question or observation you have over Acts 28. _____
_____

2. List one way Acts 28 will help you in your daily walk with God. _____
_____

3. What do you admire the most about Paul? _____
_____

4. Write out one action step you will take in your life based on today's reading. *I will* _____
_____

Request
-Devote four minutes to talking with God in prayer.

Lesson 101

Relax
-Spend one minute meditating on the words of Romans 1:16. *"For I am not ashamed of the gospel, for it is the power of God for salvation to everyone who believes, to the Jew first and also to the Greek."*

Read
-Devote four minutes to reading Romans 1.

Reflect
-Take three minutes to think through the following questions.
1. What is the most important lesson you learned from this chapter?
2. How did Paul feel about the Romans?
3. What is the importance of verse 16 to your life and the lives of others?
4. Are their similarities between the world today and the world described by Paul in verses 28-32?

Record
-Spend three minutes answering the following questions.
1. Write down one question or observation you have over this chapter. _____

2. List one thing you learned from Romans 1 that will strengthen your commitment to God. _____

3. Write down one action step you will take in your life based on today's reading. *I will* _____

Request
-Devote four minutes to talking with God.

Lesson 102

Relax
-Spend one minute in a quiet place talking with God about your life. Talk with Him as you would a close friend.

Read
-Devote four minutes to reading Romans 2.

Reflect
-Take three minutes to think through the following questions.
1. What are some important qualities we learn about God from verse 4?
2. How has God shown you kindness, tolerance and patience?
3. What impact should God's kindness have on our lives? (Compare your answer with verse 4)
4. What do we learn about the judgment day from verse 16?

Record
-Spend three minutes answering the following questions.
1. Write down one question or observation you have over this chapter. _____

2. List one way this chapter will improve your personal walk with Jesus. _____

3. Describe the condition of your spiritual heart? _____

4. Write out one action step you will take in your life based on Romans 2. *I will* _____

Request
-Devote four minutes to talking with God.

Lesson 103

Relax
   -Spend one minute in prayer thanking God for the blessings He has given you. *"If the only prayer you ever said was thank you—that would be enough!"*

Read
   -Devote four minutes to reading Psalm 75 and Psalm 76.

Reflect
   -Take three minutes to think through the following questions.
   1. What is the most important lesson you learned from these Psalms?
   2. Do you spend more time asking God for things or thanking Him for the things He has given you?
   3. Read Psalm 75:1—How can you keep the name of God near to you in your life?

Record
   -Spend three minutes answering the following questions.
   1. Write down one question or observation you have over these Psalms. _____
   2. List one way this chapter has improved your understanding of how God wants you to live your life. _____
   3. What is the most important blessing you are thankful for as a Christian? _____
   4. Write out one action step you will take in your life based on today's reading. *I will* _____

Request
   -Devote four minutes to talking with God.

Lesson 104

Relax
-Spend one minute meditating on the following statement.
"What no human being can do on his own, God will do through those who know Jesus."

Read
-Devote four minutes to reading Romans 3.

Reflect
-Take three minutes to think through the following questions.
1. What is the most interesting lesson you learned from this chapter?
2. Read verse 23—What is one thing everyone has in common?
3. How important is a relationship with Jesus in light of Romans 3:23?
4. What are some ways people try to justify their sinfulness? Are you guilty of doing this in your life?

Record
-Spend three minutes answering the following questions.
1. Write down one question or observation you have over this chapter. _____

2. List one way this chapter will strengthen your relationship with Jesus. _____

3. Write out one action step you will take in your life based on Romans 3. *I will* _____

Request
-Devote four minutes to talking with God.

Lesson 105

Relax
 -Spend one minute talking with God asking Him to help you remove any barriers in your life that are putting distance between you and Him.

Read
 -Devote four minutes to reading Romans 4.

Reflect
 -Take three minutes to think through the following questions.
 1. What is one important lesson you learned from reading this chapter?
 2. How would you define faith? (Read Hebrews 11:1)
 3. How important is faith in our relationship with Jesus?
 4. On a scale of 1-10 with 10 being powerful, how strong is your faith?

Record
 -Spend three minutes answering the following questions.
 1. Write down one question or observation you have over Romans 4. _____
 _____

 2. List one way this chapter will help you strengthen your faith in God. _____
 _____

 3. Write out one action step you will take in your life based on today's reading. *I will* _____
 _____

Request
 -Devote four minutes to strengthening your friendship with God by talking with Him in prayer.

Lesson 106

Relax
-Spend one minute meditating on the words of Romans 4:8. *"Blessed is the man whose sin the Lord will not take into account."*

Read
-Devote four minutes to reading Psalm 77.

Reflect
-Take three minutes to think through the following questions.
1. How would you describe the attitude of the Psalmist in verses 1-9?
2. Read verses 7-9—Has there ever been a time in your life when you had similar feelings?
3. Where did the writer of this Psalm find comfort?

Record
-Spend three minutes answering the following questions.
1. Write down one question or observation you have over this chapter. _____
2. List one way this chapter will help you serve others who are in need of comfort. _____
3. How will this chapter help you handle the tough times you will face in life? _____
4. Write out one action step you will take in your life based on today's reading. *I will* _____

Request
-Devote four minutes to talking with God.

Lesson 107

Relax
   -Spend one minute in a quiet place focusing your thoughts on your relationship with God.

Read
   -Devote four minutes to reading Romans 5.

Reflect
   -Take three minutes to think through the following questions.
1. What is the most important lesson you learned from this chapter?
2. Would you consider your life to be full of peace?
3. Where can we go to find peace in our life?
4. How will this chapter help you become a stronger servant for Jesus?

Record
   -Spend three minutes answering the following questions.
1. Read verses 6-10—List the words in these verses that describe someone without Jesus. _____

2. List the words or phrases in these same verses that describe a person with Jesus. _____

3. List one positive change Jesus has made in your life. _____

4. Write out one action step you will take in your life based on your reading today. *I will* _____

Request
   -Devote four minutes to talking with God thanking Him specifically for the gift of salvation.

Lesson 108

Relax
-Spend one minute meditating on the words of Romans 6:23. *"For the wages of sin is death, **but** the gift of God is eternal life in Christ Jesus our Lord."*

Read
-Devote four minutes to reading Romans 6.

Reflect
-Take three minutes to think through the following questions.
1. What is one important lesson you learned from this chapter?
2. What do we learn about the importance of baptism from this chapter?
3. What do you think of when you hear the word slave?
4. Read verses 18-19—What does it mean to be a slave of righteousness?
5. Are you currently a slave to Jesus or Satan?

Record
-Spend three minutes answering the following questions.
1. Write down one question or observation you have over this chapter. _____

2. List one way this chapter will help you in your daily walk with Jesus. _____

3. Write out one action step you will take in your life based on this chapter. *I will* _____

Request
-Devote four minutes to talking with God.

Lesson 109

Relax
-Spend one minute reading and meditating on the description of heaven found in Revelation 21:3-4.

Read
-Devote four minutes to reading Psalm 78.

Reflect
-Take three minutes to think through the following questions.
1. What is the most interesting lesson you learned from this chapter?
2. What do we learn about the importance of parents teaching their children about God from Psalm 78?
3. How would you describe God's anger?
4. Does society today see God as being a God of love or a God of wrath? How should we see God?
5. Why have people throughout time continually rebelled against God?

Record
-Spend three minutes answering the following questions.
1. Write out one question or observation you have over this chapter. _____

2. List one way this chapter has improved your overall understanding of who God is. _____

3. Write out one action step you will take in your life based on today's reading. *I will* _____

Request
-Devote four minutes to talking with God.

Lesson 110

Relax
-Spend one minute meditating on the words of Matthew 1:21. *"She will give birth to a Son, and you are to give Him the name Jesus, because He will save His people from their sins."*

Read
-Devote four minutes to reading Romans 7.

Reflect
-Take three minutes to think through the following questions.
1. What is Paul explaining with the marriage illustration he uses in this chapter?
2. Why was it difficult for the Jews to realize they had been released from the old Law?
3. Read verse 19—Does this verse ever describe you?
4. What is the most important thing Jesus has done for you?

Record
-Spend three minutes answering the following questions.
1. Write down one question or observation you have over this chapter. _____

2. List one way Romans 7 will deepen your love for Jesus. _____

3. Write down one action step you will take in your life based on today's reading. *I will* _____

Request
-Devote four minutes to talking with God.

Lesson 111

Relax
-Spend one minute meditating on the words of Romans 8:37. *"No, in all these things we are more than conquerors through Him who loved us."*

Read
-Devote four minutes to reading Romans 8.

Reflect
-Take three minutes to think through the following questions.
1. How would you describe Satan's mission?
2. What are some strategies Satan uses to try and separate you from God?
3. What is one way the Holy Spirit works in our lives as Christians?
4. What are some blessings of having a saving relationship with Jesus?

Record
-Spend three minutes answering the following questions.
1. Write out the words of the verse that impacted you the most from this chapter. _____

2. What is one lesson you learned from this chapter that will help you in your daily walk with God? ___

3. Write out one action step you will take in your life based on today's reading. *I will* _____

Request
-Devote four minutes to talking with God.

Lesson 112

Relax
-Spend one minute talking with God asking Him to be with your family and loves ones. (Job 1:5)

Read
-Devote four minutes to reading Psalm 79 and Psalm 80.

Reflect
-Take three minutes to think through the following questions.
1. What is one important lesson you learned from these Psalms?
2. In Psalm 79, the Psalmist asks several times to be returned, restored and revived. *Has there been a time in your spiritual life when you were in need of revival?*
3. Which word/phrase best describes your current relationship with God—alive *or* in need of revival?
4. How can you maintain a fire for God in your life?

Record
-Spend three minutes answering the following questions.
1. Write down one question or observation you have over these Psalms. _____

2. List one way these Psalms will help you strengthen your relationship with God. _____

3. Write out one action step you will take in your life based on today's reading. *I will* _____

Request
-Devote four minutes to talking with God.

Lesson 113

Relax
-Spend one minute meditating on the words of Psalm 37:23. *"The steps of a man are established by the Lord, and He delights in his way."*

Read
-Devote four minutes to reading Romans 9.

Reflect
-Take three minutes to think through the following questions.
1. What is the most interesting lesson you learned from this chapter?
2. Read verse 3—What do we learn about Paul from the statement he makes in this verse?
3. Describe the love Paul had for lost souls.
4. What would/could happen in the world today if Christians had Paul's love and concern for the lost?

Record
-Spend three minutes answering the following questions.
1. Write down one question or observation you have over this chapter. _____
2. List one way this chapter will help you deepen your faith in Jesus. _____
3. Write out one action step you will take in your life based on today's reading. *I will* _____

Request
-Devote four minutes to talking with God.

Lesson 114

Relax

-Spend one minute in a quiet place meditating on the words of Romans 10:15. *"How beautiful are the feet of those who bring good news of good things."*

Read

-Devote four minutes to reading Romans 10.

Reflect

-Take three minutes to think through the following questions.
1. What is one important lesson you learned from this chapter?
2. How can you strengthen your faith?
3. What did you learn from this chapter that will help you reach out to your friends and neighbors who are lost?
4. Read verse 2—Is this verse applicable to the religious world today?

Record

-Spend three minutes answering the following questions.
1. Write down one question or observation you have over this chapter. _____
2. What important fact do we learn about Jesus from verse 12? _____
3. Write out one action step you will take in your life based on today's reading. *I will* _____

Request

-Devote four minutes to talking with God.

Lesson 115

Relax
-Spend one minute in a quiet place talking with God about one area of your spiritual life you need to strengthen. *Ask God specifically for help in growing in this area!*

Read
-Devote four minutes to reading Psalm 81 and Psalm 82.

Reflect
-Take three minutes to think through the following questions.
1. What is the most interesting lesson you learned from reading these Psalms?
2. Read Psalm 8—Why is it sometimes difficult to listen to God's voice?
3. Read Psalm 81:15—Have you ever 'pretended' to live for God? What is the danger of doing this?
4. How can someone develop a 'real' relationship with God?

Record
-Spend three minutes answering the following questions.
1. Write down one question or observation you have from these Psalms. _____
2. List one way these Psalms will help you in your daily walk with God. _____
3. Write out one action step you will take in your life based on today's reading. *I will* _____

Request
-Devote four minutes to talking with God.

Lesson 116

Relax
-Spend one minute in a quiet place reading and meditating on the words of Philippians 4:8.

Read
-Devote four minutes to reading Romans 11.

Reflect
-Take three minutes to think through the following questions.
1. What is the most important lesson you learned from this chapter?
2. Read verse 22—What do we learn about God from this verse?
3. How would you put into words the unbelievable blessing of repentance God makes available to humanity?
4. What are some obstacles that sometimes prevents people from repenting and turning to God?

Record
-Spend three minutes answering the following questions.
1. Write down one question or observation you have over this chapter. _____

2. List one way this chapter will strengthen your relationship with God. _____

3. Write out one action step you will take in your life based on today's reading. *I will* _____

Request
-Devote four minutes to talking with God.

Lesson 117

Relax
-Spend one minute meditating on the words of Colossians 3:2. *"Set your mind on the things above, not on the things that are on earth."*

Read
-Devote four minutes to reading Romans 12.

Reflect
-Take three minutes to think through the following questions.
1. Read verse 1—How can you as a Christian offer your body as a living sacrifice?
2. How would you describe the values of the world?
3. In what ways is Satan trying to get you to conform to the world?
4. How can we accomplish Paul's plea in verse 2 of transforming our lives into the image of Christ?
5. What is the connection between your attitudes and your mind?

Record
-Spend three minutes answering the following questions.
1. Write down one question or observation you have over this chapter. _____

2. Read verses 4-8—List two gifts God has given you that you are currently using for Him. _____

3. Write out one action step you will take in your life based on today's reading. *I will* _____

Request
-Devote four minutes to talking with God.

Lesson 118

Relax
-Spend one minute in a quiet place talking with God about your relationships within your family, at your job and at school. *Share your concerns and struggles with God.*

Read
-Devote four minutes to reading Psalm 83 and Psalm 84.

Reflect
-Take three minutes to think through the following questions.
1. Read Psalm 84:4 and Psalm 84:10—What will you find in God's house and God's courts?
2. Have you ever felt God's 'presence' in your life?
3. Why do you think David would say that one day with God is better than a thousand days without Him?
4. Does Psalm 84:12 describe your current relationship with God?

Record
-Spend three minutes answering the following questions.
1. Write down one question or observation you have over these Psalms. _____

2. List one lesson you learned today that will help you in your daily walk with God. _____

3. Write out one action step you will take in your life based on today's reading. *I will* _____

Request
-Devote four minutes to talking with God.

Lesson 119

Relax
-Spend one minute reading and meditating on the words of Matthew 22:36-40.

Read
-Devote four minutes to reading Romans 13.

Reflect
-Take three minutes to think through the following questions.
1. What is one important lesson you learned from this chapter?
2. What is the importance of love in our relationship with God and our neighbors?
3. How do we learn to love someone?
4. How strong is your love for God?

Record
-Spend three minutes answering the following questions.
1. Write down one question or observation you have over Romans 13. _____

2. List two things that motivate you to stay away from sin. _____, _____
3. How will this chapter help you in your daily walk with God? _____

4. List one action step you will take in your life based on today's reading. *I will* _____

Request
-Devote four minutes to talking with God.

Lesson 120

Relax
-Spend one minute meditating on how it will feel to face the judgment day knowing you are saved. (1st John 5:13)

Read
-Devote four minutes to reading Romans 14.

Reflect
-Take three minutes to think through the following questions.
1. What is the most important lesson you learned from this chapter?
2. How do you envision the judgment day?
3. What did you learn about the judgment day from this chapter?
4. Read verse 8—Can you say you have completely given yourself to the Lord?

Record
-Spend three minutes answering the following question.
1. Write down one question or observation you have over this chapter. _____

2. List one way this chapter will strengthen your commitment to God. _____

3. Write down one action step you will take in your life based on today's reading. *I will* _____

Request
-Devote four minutes to talking with God.

Lesson 121

Relax
-Spend one minute meditating on the words of Psalm 86:11. *"Teach me Your way, O Lord; I will walk in Your truth; Unite my heart to fear Your name."*

Read
-Devote four minutes to reading Psalm 85 and Psalm 86.

Reflect
-Take three minutes to think through the following questions.
1. How would you describe your spiritual heart?
2. What are some things that sometimes prevent you from completely surrendering your heart to God?
3. If you wholeheartedly served God, what would that do to your ministry at school or work?
4. Is it easy or difficult for you to talk to others about God? How can you get more comfortable at it?

Record
-Spend three minutes answering the following questions.
1. Write down one question or observation you have over these chapters. _____

2. List one way these Psalms will strengthen your love for God. _____

3. Write down one action step you will take in your life based on today's reading. *I will* _____

Request
-Devote four minutes to talking with God.

Lesson 122

Relax
-Spend one minute meditating on the following quote in relationship to your responsibilities to your family. *"True leaders act with courage and stand tall in the face of adversity."*

Read
-Devote four minutes to reading Romans 15.

Reflect
-Take three minutes to think through the following questions.
1. Read verse 4—What do we learn about the purpose of the Old Testament from this verse?
2. What did you learn about leadership from Paul?
3. Do you agree with the following statement?—Paul was a great leader because he was willing to be a follower.

Record
-Spend three minutes answering the following questions.
1. Write down one question or observation you have over Romans 15. _____
2. Read verse 5—What do you need to change in your life to make your mind more like Christ's? _____
3. What is the role of prayer in leading others to Christ? _____
4. Write out one action step you will take in your life based on today's reading. *I will* _____

Request
-Devote four minutes to talking with God.

Lesson 123

Relax
-Spend one minute thinking on the following statement in relation to Philippians 4:8. *"You are today where your thoughts have brought you; you will be tomorrow where your thoughts take you."*

Read
-Devote four minutes to reading Romans 16.

Reflect
-Take three minutes to think through the following questions.
1. What is one important lesson you have learned from studying Romans?
2. What is the significance of the names mentioned by Paul in this chapter?
3. Is Paul's warning in verse 17 applicable to us today?

Record
-Spend three minutes answering the following questions.
1. Write down one question or observation you have over Romans 16. _____

2. List one way this chapter will help you strengthen your relationship with Jesus. _____

3. How does verse 20 make you feel as a Christian? __

4. Write out one action step you will take in your life based on today's reading. *I will* _____

Request
-Devote four minutes to talking with God.

Lesson 124

Relax
-Spend one minute in a quiet place meditating on the words of God in Hebrews 13:5. *"I will never desert you, nor will I ever forsake you."*

Read
-Devote four minutes to reading Psalm 87 and Psalm 88.

Reflect
-Take three minutes to think through the following questions.
1. What does the word holy mean? (Set apart)
2. How would you describe the holiness of God?
3. Are you living a life that is set apart from the world?
4. How would you describe the mood of the Psalmist from what you read in Psalm 88?

Record
-Spend three minutes answering the following questions.
1. Write down one question or observation you have over Psalm 87 and 88. _____

2. Read Psalm 88:14 and Isaiah 59:1-2—What would cause God to hide His face from you? _____

3. When God hides His face from us because of sin, has He abandoned us or have we abandoned Him? _____

4. Write out one action step you will take in your life based on today's reading. *I will* _____

Request
-Devote four minutes to talking with God.

Lesson 125

Relax
 -Spend one minute meditating on the following words while thinking of your spiritual walk. *"Our lives are a reflection of what we focus on each day."*

Read
 -Devote four minutes to reading John 1.

Reflect
 -Take three minutes to think through the following questions.
 1. What is the most interesting lesson you learned about Jesus from this chapter?
 2. How would you explain the role of John the Baptist?
 3. Read verses 41 and 42—What did Andrew do immediately after discovering Jesus was the Messiah?
 4. What group(s) of people do you have the best chance of leading to Christ?

Record
 -Spend three minutes answering the following questions.
 1. Write down one question or observation you have over John 1. _____

 2. List one way this chapter will strengthen your walk with Christ. _____

 3. Write out one action step you will take in your life based on today's reading. *I will* _____

Request
 -Devote four minutes to talking with God.

Lesson 126

Relax
- Spend one minute meditating on the following quote in relation to your service to God. *"A ship in the harbor is safe...but that's not what ships were made for."*

Read
- Devote four minutes to reading John 2.

Reflect
- Take three minutes to think through the following questions.
    1. What is your most important purpose as a Christian?
    2. Read verse 4—Why does Jesus make this statement to His mother?
    3. What effect did the signs/miracles Jesus performed have on His disciples and others?
    4. Read verse 25—Jesus had and has the power to see inside a person. *What does Jesus see when He looks in your heart?*

Record
- Spend three minutes answering the following questions.
    1. Write down one question or observation you have over this chapter. _____
    2. List one lesson you learned from this chapter that will strengthen your love for Jesus. _____
    3. Write out one action step you will take in your life based on today's reading. *I will* _____

Request
- Devote four minutes to talking with God.

Lesson 127

Relax
-Spend one minute meditating on the Psalmist's description of the nature of God's love in Psalm 89:2. *"I will declare that your love stands firm forever."*

Read
-Devote four minutes to reading Psalm 89.

Reflect
-Take three minutes to think through the following questions.
1. If you had to describe God with one word, what word would you choose?
2. What are some characteristics we learn about God from this Psalm? (vs. 2, 5, 8, 16, 38, 46)
3. What do we learn about the wrath of God from this Psalm?
4. What are some things David did in his life to incur the wrath of God?

Record
-Spend three minutes answering the following questions.
1. Write down one question or observation you have over this Psalm. _____

2. What did you learn about God from this Psalm that will help you in your daily spiritual walk? _____

3. List one action step you will take in your life based on today's reading. *I will* _____

Request
-Devote four minutes to talking with God.

Lesson 128

Relax
 -Spend one minute meditating on the words of John 3:16.
 "For God so loved the world that He gave His one and only Son that whosoever believes in Him shall not perish but have eternal life."

Read
 -Devote four minutes to reading John 3.

Reflect
 -Take three minutes to think through the following questions.
   1. How would you describe Nicodemus?
   2. Read verses 19-20—Do these verses accurately describe today's society?
   3. What is the importance of belief in our relationship with Jesus?
   4. How would you describe God's relationship with His Son, Jesus?

Record
 -Spend three minutes answering the following questions.
   1. Write down one question or observation you have over John 3. _____

   2. List one way this chapter will strengthen your relationship with Jesus. _____

   3. Write out one action step you will take in your life based on today's reading. *I will* _____

Request
 -Devote four minutes to talking with God.

Lesson 129

Relax
-Spend one minute meditating on the following quote. *"Go over, go under, go around or go through. But never give up."*

Read
-Devote four minutes to reading John 4.

Reflect
-Take three minutes to think through the following questions.
1. How would you describe the woman that Jesus met at the well?
2. What is one lesson you learned about ministering to others from Jesus' interaction with this woman?
3. Read verses 23-24—How can we worship God in spirit and in truth?
4. Read verse 35—Is Jesus' statement about the fields ripe for harvest applicable to today's society?

Record
-Spend three minutes answering the following questions.
1. Write down one question or observation you have over this chapter. _____

2. In one sentence, describe Jesus' mission while on earth. _____

3. Write out one action step you will take in your life based on today's reading. *I will* _____

Request
-Devote four minutes to talking with God.

Lesson 130

Relax
-Spend one minute meditating on these words found in Psalm 90:2. *"Even from everlasting to everlasting, you are God."*

Read
-Devote four minutes to reading Psalm 90 and Psalm 91.

Reflect
-Take three minutes to think through the following questions.
1. Are you looking forward to the return of the Lord?
2. A thousand years seems like a long, long time. *What is a thousand years to God?*
3. Read Psalm 90:12—Why is it important that we learn to number our days?

Record
-Spend three minutes answering the following questions.
1. Write out one question or observation you have over these Psalms. _____

2. List one way these Psalms will help you strengthen your relationship with God. _____

3. List one reason why you trust in God. _____

4. Write out one action step you will take in your life based on today's reading. *I will* _____

Request
-Devote four minutes to talking with God asking specifically for strength in leading your family spiritually.

Lesson 131

Relax
   -Spend one minute in a quiet place focusing on the holiness of God.

Read
   -Devote four minutes to reading John 5.

Reflect
   -Take three minutes to think through the following questions.
1. Why does Jesus ask this man if he wants to get well in verse 6?
2. Can you think of any possible reason why this man would not want to be healed?
3. How would his life change if he were healed?
4. In what ways are we in the same condition as this sick man? *We need to be healed by Jesus.*
5. Does the statement *with privilege comes responsibility* relate to us as Christians?

Record
   -Spend three minutes answering the following questions.
1. List one way our lives should be different after becoming a Christian. _____
2. Think back to the day you were baptized—How did your life change after being baptized? _____
3. Write down one action step you will take in your life based on today's reading. *I will* _____

Request
   -Devote four minutes to talking with God.

Lesson 132

Relax
 -Spend one minute meditating on the most amazing part of Jesus' life. *What amazes you the most about Jesus?*

Read
 -Devote four minutes to reading John 6.

Reflect
 -Take three minutes to think through the following questions.
 1. What impact did the miracle of feeding the 5000 with only 5 loaves and 2 fish have on the people?
 2. What expectation did many people have concerning Jesus? (verse 15)
 3. How would you have responded if you had been in the boat with the apostles and witnessed Jesus walking on the water?
 4. Many of Jesus' disciples deserted Him in this chapter. *Why do people turn away from Jesus today?*

Record
 -Spend three minutes answering the following questions.
 1. Write down one question or observation you have from this chapter. _____
 2. List one reason why you should stay committed to Jesus. _____
 3. Write out one action step you will take in your life based on today's reading. *I will* _____

Request
 -Devote four minutes to talking with God.

Lesson 133

Relax
-Spend one minute meditating on the words of Psalm 103:12. *"As far as the east is from the west, So far has He removed our sins from us."*

Read
-Devote four minutes to reading Psalm 92 and Psalm 93.

Reflect
-Take three minutes to think through the following questions.
1. Which verse from Psalm 92 discusses the Psalmist's daily relationship with God?
2. Why is it important to have a daily walk with God?
3. What are some things preventing you from having a daily relationship with God?
4. Describe the benefits of living righteous in the eyes of God. (Psalm 92:12-13)

Record
-Spend three minutes answering the following questions.
1. Write down one question or observation you have over these Psalms. _____
2. List one lesson you learned from these Psalms that help you develop a daily walk with God. _____
3. Write down one action step you will take in your life based on today's reading. *I will* _____

Request
-Devote four minutes to talking with God.

Lesson 134

Relax
-Spend one minute meditating on the following quote. *Our lives are not determined by what happens to us, but how we react to what happens.*

Read
-Devote four minutes to reading John 7.

Reflect
-Take three minutes to think through the following questions.
1. Read verse 7—How was Jesus' life affected by His stance against evil?
2. Is it easy or difficult for you to stand up for God?
3. What are some possible struggles you may face when you stand up for God?
4. Read the relax quote in the above section again. *How do you react to the opportunities you have to stand up for God?*

Record
-Spend three minutes answering the following questions.
1. Write down one question or observation you have over John 7. _____

2. List one lesson you learned from Jesus in this chapter that will help you stand up for Him. _____

3. Write out one action step you will take in your life based on today's reading. *I will* _____

Request
-Devote four minutes to talking with God.

Lesson 135

Relax
-Spend one minute meditating on the words of Jesus in John 8:32 as you think about the freedom Jesus gives from sin.

Read
-Devote four minutes to reading John 8.

Reflect
-Take three minutes to think through the following questions.
1. How would you describe the attitude of Jesus towards this woman in verses 1-11?
2. Was Jesus justifying her past sinfulness by stepping in on her behalf?
3. What is the significance of Jesus' words in verse 11?
4. What are some things you have in common with the woman caught in adultery?

Record
-Spend three minutes answering the following questions.
1. Write down one question or observation you have from this chapter. _____
2. Why were the Jews seeking to kill Jesus in the latter part of this chapter? _____
3. List one way this chapter will deepen your understanding of who God is. _____
4. Write out one action step you will take in your life based on today's reading. *I will* _____

Request
-Devote four minutes to talking with God.

Lesson 136

Relax
    -Spend one minute in a quiet place talking with God about your life. *Take time to thank Him for your blessings and ask for strength to follow in His Son's footsteps.*

Read
    -Devote four minutes to reading Psalm 94 and Psalm 95.

Reflect
    -Take three minutes to think through the following questions.
1. Read Psalm 94:11—How do you feel knowing God knows your every thought?
2. Which verse from Psalm 94 brings you the most comfort as a Christian? Why?
3. Why should you want to worship and kneel before God?
4. Read the questions from Psalm 94:16—Are these important questions for us today? Why?

Record
    -Spend three minutes answering the following questions.
1. Write down one question or observation you have over these Psalms. _____

2. List one lesson you learned from these Psalms that will strengthen your daily walk with God. _____

3. Write out one action step you will take in your life based on today's reading. *I will* _____

Request
    -Devote four minutes to talking with God.

Lesson 137

Relax
    -Spend one minute meditating on the words found in Psalm 94:14. *"For the Lord will not abandon His people, nor will He forsake His inheritance."*

Read
    -Devote four minutes to reading John 9.

Reflect
    -Take three minutes to think through the following questions.
        1. What is the most important lesson you learned from this chapter?
        2. How would you describe the attitude of the Pharisees toward Jesus?
        3. What were some obstacles preventing the Pharisees from accepting Jesus as the Messiah?
        4. Who had more faith in this chapter, the blind man or his parents? How do you know?

Record
    -Spend three minutes answering the following questions.
        1. Write down one question or observation you have over this chapter. _____

        2. List one way this chapter will strengthen your daily walk with Jesus. _____

        3. Write out one action step you will take in your life based on today's reading. *I will* _____

Request
    -Devote four minutes to talking with God.

Lesson 138

Relax
-Spend one minute meditating on your life in relation to the promise Jesus gives in John 10:10. *"I came that they may life and have it abundantly."*

Read
-Devote four minutes to reading John 10.

Reflect
-Take three minutes to think through the following questions.
1. Do you agree with the following statement? *Christianity is the best life to live! Why or why not?*
2. What are some qualities we learn about Jesus from this chapter that should motivate us to live for Him?
3. What analogy does Jesus use to show His love for us as Christians?
4. Read verse 30—What was the significance of this statement for Jesus' life and for ours today?

Record
-Spend three minutes answering the following questions.
1. Write down one question or observation you have over this chapter. _____
_____
2. What excites you most about being a Christian? ___
_____
3. Write out one action step you will take in your life based on today's reading. *I will* _____
_____

Request
-Devote four minutes to talking with God.

Lesson 139

Relax
    -Spend one minute meditating on the greatness of the Lord as you read Psalm 96:4. *"For great is the Lord and greatly to be praised."*

Read
    -Devote four minutes to reading Psalm 96 and Psalm 97.

Reflect
    -Take three minutes to think through the following questions.
1. What is the most important lesson you learned about God from these two Psalms?
2. What lessons do we learn from Psalm 96 that should motivate us to tell others about God?
3. How should we as Christians feel about evil?
4. How do you feel about the coming judgment day?

Record
    -Spend three minutes answering the following questions.
1. Write down one question or observation you have from these two Psalms. _____
2. List one way these Psalms will help you in your ministry to others. _____
3. List one blessing you need to thank God for giving you. _____
4. Write out one action step you will take in your life based on today's reading. *I will* _____

Request
    -Devote four minutes to talking with God.

Lesson 140

Relax
-Spend one minute meditating on the compassion Jesus has for you and others while reading the simple, but powerful words found in John 11:35, *"Jesus wept."*

Read
-Devote four minutes to reading John 11.

Reflect
-Take three minutes to think through the following questions.
1. Is it easier for you to see Jesus as being God or man?
2. What did you learn from this chapter that reveals the human side of Jesus? What about the God side?
3. In what ways are we in the same condition spiritually as Lazarus was physically?
4. What impact does Jesus' raising of Lazarus from the dead have on your life?

Record
-Spend three minutes answering the following questions.
1. Write down one question or observation you have over this chapter. _____

2. List one lesson you learned in this chapter that will strengthen your relationship with Jesus. _____

3. Write out one action step you will take in your life based on today's reading. *I will* _____

Request
-Devote four minutes to talking with God.

Lesson 141

Relax
    -Spend one minute meditating on the following words as you look forward to heaven. *Hold to the hope, the belief, the conviction that there is a better life, a better world beyond the horizon.*

Read
    -Devote four minutes to reading John 12.

Reflect
    -Take three minutes to think through the following questions.
    1. Read verse 27—What does this verse reveal to us about Jesus?
    2. Did Jesus want to suffer and die on the cross? What motivated Jesus to go through with this?
    3. What is your purpose as a Christian?
    4. What do we learn from Jesus that will help us fulfill our responsibilities to God as His children?

Record
    -Spend three minutes answering the following questions.
    1. Write down one question or observation you have over this chapter. _____
    2. List one way this chapter will strengthen your commitment to God. _____
    3. Write out one action step you will take in your life based on today's reading. *I will* _____

Request
    -Devote four minutes to talking with God.

Lesson 142

Relax
-Spend one minute in a quiet place focusing your thoughts on God. *Talk with your Father about a struggle you are facing in your Christian walk.*

Read
-Devote four minutes to reading Psalm 98 – Psalm 101.

Reflect
-Take three minutes to think through the following questions.
1. What is the most important lesson you learned from these Psalms?
2. How would you describe the greatness of God?
3. Read Psalm 101:3a—Are you accomplishing this in your daily life?
4. What is your greatest struggle in striving to live for Christ?

Record
-Spend three minutes answering the following questions.
1. Write down one question or observation you have over these Psalms. _____

2. List one way these Psalms have deepened your understanding of who God is. _____

3. Write out one action step you will take in your life based on today's reading. *I will* _____
   _____

Request
-Devote four minutes to talking with God.

Lesson 143

Relax
-Spend one minute meditating on the words of Jesus found in John 13:34. *"A new commandment I give to you, that you love one another, even as I have loved you."*

Read
-Devote four minutes to reading John 13.

Reflect
-Take three minutes to think through the following questions.
1. Why did Jesus wash the feet of His disciples?
2. Do you think Jesus washed Judas' feet even while knowing he would soon betray Him?
3. Read verse 37—How did Peter go from making this statement to later denying ever knowing Jesus?
4. Compare the love Jesus demonstrated for others with the love you currently show people.

Record
-Spend three minutes answering the following questions.
1. Write down one question or observation you have from today's reading. _____
2. How will the world around us (our friends, relatives, neighbors) know we are disciples of Jesus? _____
3. Write out one action step you will take in your life based on today's reading. *I will* _____

Request
-Devote four minutes to talking with God.

Lesson 144

Relax
  -Spend one minute meditating on Jesus' words in John 14:6. *"I am the way the truth and the life. No one comes to the Father except through me."*

Read
  -Devote four minutes to reading John 14.

Reflect
  -Take three minutes to think through the following questions.
  1. Read verse 5—How would you describe Thomas based on this question he asks Jesus?
  2. What did you learn about heaven from verse 2?
  3. Read verse 8—Why does Philip make this statement to Jesus? What does this reveal to us about Philip?
  4. What was the role of the Holy Spirit in the future ministry efforts of the apostles?

Record
  -Spend three minutes answering the following questions.
  1. Write down one question or observation you have over this chapter. _____
  2. Describe the connection between love, friendship and obedience in our relationship with Jesus. _____
  3. Write out one action step you will take in your life based on today's reading. *I will* _____

Request
  -Devote four minutes to talking with God.

Lesson 145

Relax
-Spend one minute meditating on the connection between the following quote and our Christian journey. *"Two roads diverged in a wood and I—I took the one less traveled by, and that has made all the difference."*

Read
-Devote four minutes to reading Psalm 102.

Reflect
-Take three minutes to think through the following questions.
1. How would you describe the mood of the Psalmist as reflected by the words of this Psalm?
2. Read verse 18—How can you help your future descendants have a faithful relationship with God?
3. Compare our time here on earth verses eternity.
4. Read verse 15—Do we as a nation fear God? Does your answer to this question scare you?

Record
-Spend three minutes answering the following questions.
1. Write down one question or observation you have over this Psalm. _____

2. List one characteristic you learned about God in this Psalm that will strengthen your love for Him. _____

3. Write out one action step you will take in your life based on today's reading. *I will* _____

Request
-Devote four minutes to talking with God.

Lesson 146

Relax
    -Spend one minute meditating on the following words from Psalm 102:27. *"But you (God) remain the same, and your years will never end."*

Read
    -Devote four minutes to reading John 15.

Reflect
    -Take three minutes to think through the following questions.
1. What is the importance of bearing fruit as a Christian?
2. In what ways are you fruitful as a Christian?
3. What is one area of ministry where you can be more productive as a Christian?
4. How can you remain in the love of Jesus?

Record
    -Spend three minutes answering the following questions.
1. Write down one question or observation you have over this chapter. _____

2. What is the ultimate test of love? _____

3. How do you feel knowing Jesus did this for you? __

4. Write out one action step you will take in your life based on today's reading. *I will* _____

Request
    -Devote four minutes to talking with God.

Lesson 147

Relax
-Spend one minute meditating on the following quote in relation to our Christian journey. *"Success requires persistence. If it is the right thing, never, never, give up."*

Read
-Devote four minutes to reading John 16.

Reflect
-Take three minutes to think through the following questions.
1. What is the most interesting lesson you learned from this chapter?
2. Why did Jesus leave His ministry, that He ultimately died for, in the hands of ordinary men?
3. How did Jesus prepare His apostles for their upcoming ministry?
4. How would you describe the relationship between being in Christ and being at peace?

Record
-Spend three minutes answering the following questions.
1. Write down one question or observation you have over this chapter. _____

2. How can we develop persistence as a Christian? ___

3. Write out one action step you will take in your life based on today's reading. *I will* _____

Request
-Devote four minutes to talking with God.

Lesson 148

Relax
-Spend one minute in a quiet place focusing your thoughts and mind on God as you talk to your Dad (God) about your life.

Read
-Devote four minutes to reading Psalm 103.

Reflect
-Take three minutes to think through the following questions.
1. What is the most amazing blessing about being a Christian?
2. How would you describe your current relationship with God?
3. What is one thing you can do to strengthen your relationship with God?
4. Which verse impacted you the most from today's reading?

Record
-Spend three minutes answering the following questions.
1. Write down one question or observation you have over today's reading. _____
_____

2. How is God's love for you described in this Psalm?
_____
_____

3. Write out one action step you will take in your life based on today's reading. *I will* _____
_____

Request
-Devote four minutes to talking with God.

Lesson 149

Relax
-Spend one minute in a quiet place meditating on the following statement. *As Jesus reigns at the right hand of God, He prays for you!*

Read
-Devote four minutes to reading John 17.

Reflect
-Take three minutes to think through the following questions.
1. How can you bring God glory with your life?
2. How did Jesus glorify God with His life?
3. Is there such a thing as absolute truth/absolute right and wrong?
4. Read verse 17—Where can we go to find truth?

Record
-Spend three minutes answering the following questions.
1. Write down one question or observation you have over this chapter. _____

2. List one thing Jesus prayed for concerning His apostles. _____

3. What is one lesson you learned about prayer from Jesus in today's reading? _____

4. Write out one action step you will take in your life based on today's reading. *I will* _____

Request
-Devote four minutes to talking with God.

Lesson 150

Relax
-Spend one minute meditating on the following statement.
"Jesus enveloped His most important decisions with prayer. So should we!"

Read
-Devote four minutes to reading John 18.

Reflect
-Take three minutes to think through the following questions.
1. What was Jesus doing immediately before He was arrested?
2. How did Jesus use prayer in His life? How do you use prayer?
3. How would you describe Peter based on what you read about him in this chapter?
4. What is one thing you have in common with Peter?

Record
-Spend three minutes answering the following questions.
1. Write down one question or observation you have over this chapter. _____

2. Read verse 38—What does the world today have in common with Pilate? _____

3. Write out one action step you will take in your life based on today's reading. *I will* _____

Request
-Devote four minutes to talking with God asking specifically for help with a tough decision you may be facing.

Lesson 151

Relax
    -Spend one minute in meditation rejoicing in the power and glory of the Lord. *"May my meditation be pleasing to Him, as I rejoice in the Lord." Psalm 104:34*

Read
    -Devote four minutes to reading Psalm 104.

Reflect
    -Take three minutes to think through the following questions.
        1. What is the most interesting lesson you learned from this Psalm?
        2. How would you describe the power of God?
        3. Are we sometimes as humans guilty of placing limits on the power of God?
        4. Is it difficult for you to give complete control of your life to God? Why or why not?

Read
    -Spend three minutes answering the following questions.
        1. Write down one question or observation you have over this Psalm. _____

        2. List one way this chapter will strengthen your daily walk with God. _____

        3. Write out one action step you will take in your life based on today's reading. *I will* _____

Request
    -Devote four minutes to talking with God.

Lesson 152

Relax
    -Spend one minute meditating on the following quote.
    "Obstacles are those frightful things you see when you take your eyes off the goal."

Read
    -Devote four minutes to reading John 19.

Reflect
    -Take three minutes to think through the following questions.
1. How do you feel when you reflect on all the pain and suffering Jesus went through in His death?
2. How do you feel when you realize He suffered all of these horrible things for you?
3. Do you think Pilate believed in Jesus as the Son of God? Why or why not?
4. Read verse 10—Why didn't Pilate use his power to free Jesus?

Record
    -Spend three minutes answering the following questions.
1. Write down one question or observation you have over this chapter. _____
2. List one lesson you learned from today's reading that will strengthen your love for Jesus. _____
3. Write out one action step you will take in your life based on today's reading. *I will* _____

Request
    -Devote four minutes to talking with God.

Lesson 153

Relax
-Spend one minute in a quiet place focusing your thoughts and mind on God.

Read
-Devote four minutes to reading John 20.

Reflect
-Take three minutes to think through the following questions.
1. What is the significance of Jesus' resurrection for Christians?
2. Read verse 30—Imagine what it would have been like to have witnessed all of Jesus' miracles!
3. Does society today have anything in common with the Thomas we see in this chapter?
4. What is the importance of belief in our relationship with Jesus? (John 8:24)

Record
-Spend three minutes answering the following questions.
1. Write down one question or observation you have over this chapter. _____

2. Why did John record many of the miraculous signs performed by Jesus in his account of Jesus' life? ___

3. Write out one action step you will take in your life based on today's reading. *I will* _____

Request
-Devote four minutes to talking with God.

Lesson 154

Relax

-Spend one minute meditating on the following quote in relation to our purpose as Christians. *"No light shines brighter than when it shines in the darkness."*

Read

-Devote four minutes to reading John 21.

Reflect

-Take three minutes to think through the following questions.
1. What was Peter doing when Jesus first called Him to follow Him? (Matthew 4:18-20)
2. Where did Jesus have to go to find Peter after His death and resurrection?
3. Why do you think Peter returned to his life as a fisherman?
4. Why do many people today desert Christianity and return to the world?

Record

-Devote three minutes to answering the following questions.
1. Write down one question or observation you have over this chapter. _____

2. Describe Peter's relationship with Jesus. _____

3. Describe your relationship with Jesus. _____

4. Write out one action step you will take in your life based on today's reading. *I will* _____

Request

-Spend four minutes in prayer talking with your Father.

Lesson 155

Relax
-Spend one minute meditating on the words of Psalm 105:4.
"Seek the Lord and His strength; seek His face continually."

Read
-Devote four minutes to reading Psalm 105.

Reflect
-Take three minutes to think through the following questions.
1. How has God proven His faithfulness to you in your life?
2. What are you doing in your life right now to "seek God's face continually?"
3. Read Matthew 7:7-8—What will happen in our lives if we seek God?
4. What is the connection between Psalm 105:45 and seeking God?

Record
-Spend three minutes answering the following questions.
1. Write down one question or observation you have over this Psalm. _____
2. List one lesson you learned from this Psalm that will strengthen your relationship with God. _____
3. Write out one action step you will take in your life based on today's reading. *I will* _____

Request
-Devote four minutes to talking with God. *Write down two requests you are currently making to God.*

Lesson 156

Relax
-Spend one minute meditating on the following words. *"The soul is dyed the color of its thoughts. Think only on those things that are in line with God's principles."*

Read
-Devote four minutes to reading Psalm 106.

Reflect
-Take three minutes to think through the following questions.
1. How would you describe Israel's relationship with God based on what you read in this Psalm?
2. Read verse 21—How could the Israelites forget God after all the things He had done for them?
3. Have you ever been guilty of forgetting God?
4. What are some results of failing to listen to and obey the voice of God?

Record
-Spend three minutes answering the following questions.
1. Write down one question or observation you have over this Psalm. _____

2. List one lesson you learned from the Israelites in this Psalm that will strengthen your walk with God. ____

3. Write out one action step you will take in your life based on today's reading. *I will* _____

Request
-Devote four minutes to talking with God.

Lesson 157

Relax
-Spend one minute in a quiet place meditating on your personal relationship with God. *Ask God for strength and wisdom to seek Him daily!*

Read
-Devote four minutes to reading 1st Corinthians 1.

Reflect
-Take three minutes to think through the following questions.
1. How do you think these people felt when they received this letter from Paul?
2. What message from this first chapter of Corinthians is most relevant to your spiritual life?
3. Why was the message of the cross a stumbling block to the Jews and foolishness to the Gentiles?

Record
-Spend three minutes answering the following questions.
1. Write out one question or observation you have over this chapter. _____

2. How can Christians today achieve unity? _____

3. List one way this chapter will strengthen your daily walk with Jesus. _____

4. Write out one action step you will take in your life based on today's reading. *I will* _____

Request
-Devote four minutes to talking with God.

Lesson 158

Relax
-Spend one minute meditating on your attitude towards God and your family while thinking on the following words.
"Your attitude determines your altitude in life."

Read
-Devote four minutes to reading Psalm 107.

Reflect
-Take three minutes to think through the following questions.
1. Why do people rebel against the Lord?
2. Have you ever been guilty of rebelling against God? What motivated you to come back to God?
3. Is there such a thing as good rebellion? (Read Daniel 3:16-18)
4. How would you describe your current attitude towards God and the church?

Record
-Spend three minutes answering the following questions.
1. Write down one question or observation you have over this chapter. _____
2. List one lesson you learned about God from this Psalm that will prevent you from rebelling against Him? _____
3. Write out one action step you will take in your life based on today's reading. *I will* _____

Request
-Devote four minutes to talking with God.

Lesson 159

Relax
- Spend one minute meditating on the words of 1st Corinthians 2:9. *"No eye has seen, no ear has heard, no mind has conceived what God has prepared for those who love Him."*

Read
- Devote four minutes to reading 1st Corinthians 2.

Reflect
- Take three minutes to think through the following questions.
    1. How does 1st Corinthians 2:9 impact your life as a Christian?
    2. What is the importance of our mind in relation to the choices we make?
    3. *"Garbage In, Garbage Out!"* Are you filling your mind with garbage or with the Spirit?
    4. Read verse 16—How can you develop a Christ-like mind?

Record
- Spend three minutes answering the following questions.
    1. Write down one question or observation you have over this chapter. _____
    2. List one way this chapter will strengthen your daily walk with Jesus. _____
    3. Write out one action step you will take in your life based on today's reading. *I will* _____

Request
- Devote four minutes to talking with God.

Lesson 160

Relax
-Spend one meditating on the qualities of God revealed in Psalm 108:4. *"For your lovingkindness is great above the heavens, and your truth reaches to the skies."*

Read
-Devote four minutes to reading Psalm 108 and Psalm 109.

Reflect
-Take three minutes to think through the following questions.
1. What is the most important lesson you learned from each of these Psalms?
2. Read Psalm 108:1—How can you develop a heart that is steadfast/committed to God?
3. How would you describe the mood of the Psalmist in Psalm 109? Have you ever felt like this?
4. Read Psalm 109:4—What is the significance of this verse to your life?

Record
-Spend three minutes answering the following questions.
1. Write down one question or observation you have over these Psalms. _____
2. List one way these Psalms will deepen your love for God. _____
3. Write out one action step you will take in your life based on today's reading. *I will* _____

Request
-Devote four minutes to talking with God.

Lesson 161

Relax
 -Spend one minute meditating on the words of 1st Corinthians 3:9. *"For we are God's fellow workers; you are God's field, God's building."*

Read
 -Devote four minutes to reading 1st Corinthians 3.

Reflect
 -Take three minutes to think through the following questions.
 1. How does verse 9 in this chapter impact you as a Christian?
 2. What is the importance of a building's foundation?
 3. What are some different foundations people can build their lives on?
 4. What foundation are you building your life on?
 5. Read verses 16-17—How do these verses impact you and how you are currently treating your body?

Record
 -Spend three minutes answering the following questions.
 1. Write down one question or observation you have over this chapter. _____

 2. List two ways you are growing as a Christian. _____, _____

 3. Write out one action step you will take in your life based on today's reading. *I will* _____

Request
 -Devote four minutes to talking with God.

Lesson 162

Relax
    -Spend one minute in a quiet place meditating on God's name and what He means to you in your life. *"Holy and awesome is His (God's) name." Psalm 111:9*

Read
    -Devote four minutes to reading Psalm 110 – Psalm 114.

Reflect
    -Take three minutes to think through the following questions.
1. If you had to describe the name of God with only one word, what word would you choose? Why?
2. What does it mean to be wise? How can one grow in wisdom?
3. Read Psalm 111:10—What is the beginning point of wisdom according to the Psalmist?
4. Do you think people today fear the Lord? Why or why not? Do you fear the Lord?

Record
    -Spend three minutes answering the following questions.
1. Write down one question or observation you have over these Psalms. _____

2. List one lesson you learned about God from these Psalms that will strengthen your walk with Him. ___

3. Write out one action step you will take in your life based on today's reading. *I will* _____

Request
    -Devote four minutes to talking with God.

Lesson 163

Relax
    -Spend one minute in a quiet place talking with God about your life. *Share with Him one goal you are working towards and one struggle you are currently facing.*

Read
    -Devote four minutes to reading 1st Corinthians 4.

Reflect
    -Take three minutes to think through the following questions.
1. Read verse 16—Why does Paul make this statement to the Corinthians?
2. Do you think Paul is arrogant in making such a bold statement? Why or why not?
3. What is the significance of Paul's statement in verse 16 to your Christian walk?
4. Think on the following words: *You may be the only Bible your friends will ever read!*

Record
    -Spend three minutes answering the following questions.
1. Write down one question or observation you have over this chapter. _____

2. List one lesson you learned from this chapter that will strengthen your daily walk with Jesus. _____

3. Write out one action step you will take in your life based on today's reading. *I will* _____

Request
    -Devote four minutes to talking with God.

Lesson 164

Relax

-Spend one minute in a quiet place meditating on the words of Psalm 116:15. *"Precious in the sight of the Lord is the death of His saints."*

Read

-Devote four minutes to reading Psalm 115 – Psalm 117.

Reflect

-Take three minutes to think through the following questions.
1. What is the most important lesson you learned from today's reading?
2. Why should you put your trust in the Lord?
3. What are some things you are passionate about in your life?
4. Are you passionate about Jesus? How can you become more passionate about Jesus?

Record

-Spend three minutes answering the following questions.
1. Write down one question or observation you have over these Psalms. _____
   _____
2. List one way these Psalms will strengthen your daily relationship with God. _____
   _____
3. Write out one action step you will take in your life based on today's reading. *I will* _____
   _____

Request

-Devote four minutes to talking with God.

Lesson 165

Relax
   -Spend one minute meditating on the following quote. *"We must adjust to an ever changing road...while holding on to God's unchanging principles."*

Read
   -Devote four minutes to reading 1st Corinthians 5 and 6.

Reflect
   -Take three minutes to think through the following questions.
     1. What problem was Paul addressing in chapter 5?
     2. How do our friends impact our actions?
     3. Which verse in today's reading clearly condemns homosexuality?
     4. What are some consequences of becoming involved in sexual immorality?

Record
   -Spend three minutes answering the following questions.
     1. Write down one question or observation you have over this chapter. _____

     2. How do you overcome the temptations you face? _____

     3. What did you learn from 1st Corinthians 6:18-20 that will help you overcome sexual temptation?____

     4. Write out one action step you will take in your life based on today's reading. *I will* _____

Request
   -Devote four minutes to talking with God.

Lesson 166

Relax
  -Spend one minute meditating on the words of Psalm 118:24. *"This is the day which the Lord has made; let us rejoice and be glad in it."*

Read
  -Devote four minutes to reading Psalm 118.

Reflect
  -Take three minutes to think through the following questions.
  1. In your expert opinion, what is the main theme of Psalm 118?
  2. How many times does the phrase "His lovingkindness is everlasting" appear in this Psalm?
  3. What is the importance of love in our relationship with God?
  4. Would you describe your love for God as everlasting? Why or why not?

Record
  -Spend three minutes answering the following questions.
  1. Write down one question or observation you have over this Psalm. _____
  2. List one lesson you learned from Psalm 118 that will strengthen your love for God. _____
  3. Write out one action step you will take in your life based on today's reading. *I will* _____

Reflect
  -Devote four minutes to talking with God.

Lesson 167

Relax
-Spend one minute meditating on your relationship with your family. *Ask God for strength to become a better husband, wife, son, daughter, etc.*

Read
-Devote four minutes to reading 1st Corinthians 7.

Reflect
-Take three minutes to think through the following questions.
1. What is the most important quality you are looking for *(or looked for)* in a future spouse? Why?
2. What is the most important lesson you learned about marriage from this chapter?
3. Read verses 10-11—What do we learn about divorce and remarriage from these verses?
4. Is the world today following these teachings?

Record
-Spend three minutes answering the following
1. Write down one question or observation you have over this chapter. _____

2. List one lesson you learned from this chapter that will strengthen your relationship with your family.
_____

3. Write out one action step you will take in your life based on today's reading. *I will* _____

Request
-Devote four minutes to talking with God.

Lesson 168

Relax
-Spend one minute meditating on the following thought.
"God has the answers for every question, struggle and moral dilemma you encounter!"

Read
-Devote four minutes to reading Psalm 119:1-64.

Reflect
-Take three minutes to think through the following questions.
1. How has your attitude about God's Word changed over the years?
2. Read verse 9—What is the connection between purity and God's Word?
3. When you are facing a tough choice where do you turn for answers? Do you turn to God?
4. Which verse from today's reading impacts you the most?

Record
-Spend three minutes answering the following questions.
1. Write down one question or observation you have over today's reading. _____

2. List one lesson you learned from today's study that will help you *seek God with all your heart*. _____

3. Write out one action step you will take in your life based on today's reading. *I will* _____

Request
-Devote four minutes to talking with God.

Lesson 169

Relax
-Spend one minute in a quiet place meditating on the words of 1st Corinthians 9:24 as you think about your Christian race. *"Run in such a way that you may win."*

Read
-Devote four minutes to reading 1st Corinthians 8 and 9.

Reflect
-Take three minutes to think through the following questions.
1. What responsibility do we have as Christians to others as described by Paul in chapter 8?
2. How can we prevent ourselves from becoming a stumbling block to others?
3. In what ways is Christianity similar to a race?
4. Would you win the prize of eternal life if your race ended today? Why or why not?

Record
-Spend three minutes answering the following questions.
1. Write down one question or observation you have over these chapters. _____
_____

2. List one area of spiritual fitness you need to strengthen in your life in order to better run the Christian race. _____
_____

3. Write out one action step you will take in your life based on today's reading. *I will* _____
_____

Request
-Devote four minutes to talking with God.

Lesson 170

Relax
-Spend one minute meditating on the words of Psalm 119:105. *"Your word is a lamp to my feet and a light to my path."*

Read
-Devote four minutes to reading Psalm 119:65-120.

Reflect
-Take three minutes to think through the following questions.
1. What is the significance of the statement David made in verse 94?
2. Can you honestly say to God, "I am Yours?" What does it take to be able to make this statement?
3. Read verse 115—How do you know who's OK to hang out with and who isn't?
4. Think about your friends—Do they encourage you to love and follow God? If not, what should you do?

Record
-Spend three minutes answering the following questions.
1. Write down one question or observation you have over these verses. _____

2. List one way these verses have strengthened your desire to pursue a relationship with God. _____

3. Write out one action step you will take in your life based on today's reading. *I will* _____

Request
-Devote four minutes to talking with God.

Lesson 171

Relax
-Spend one minute meditating on the following words. *"God is faithful; He will not let you be tempted beyond what you can bear."*

Read
-Devote four minutes to reading 1st Corinthians 10.

Reflect
-Take three minutes to think through the following questions.
1. What are some lessons we can learn from the mistakes of the Israelites?
2. What are some lessons you have learned from your own mistakes? Would you rather learn from your mistakes or the mistakes of others?
3. Do you really believe the words of 1st Corinthians 10:13?
4. Reflect on a time in your life when God provided you escape from a temptation.

Record
-Spend three minutes answering the following questions.
1. Write down one question or observation you have over this chapter. _____
2. List one temptation you are currently facing, and how this chapter will help you overcome it. _____
3. Write down one action step you will take in your life based on today's reading. *I will* _____

Request
-Devote four minutes to talking with God.

Lesson 172

Relax
-Spend one minute meditating on words of Psalm 119:103. *"How sweet are Your words to my taste! Yes sweeter than honey to my mouth."*

Read
-Devote four minutes to reading Psalm 119:121-176

Reflect
-Take three minutes to think through the following questions.
1. How would you describe the commandments of God?
2. Read verse 133—Have you ever felt as if sin had control over your life?
3. How can you prevent sin (iniquity) from having control over you?
4. What are some obstacles that prevent you from spending more time studying God's Word?

Record
-Spend three minutes answering the following questions.
1. Write down one question or observation you have over these verses. _____
2. List two verses from today's reading that support the fact there are absolute/everlasting truths to live by. _
3. Write out one action step you will take in your life based on today's reading. *I will* _____

Request
-Devote four minutes to talking with God.

Lesson 173

Relax
-Spend one minute in a quiet place talking with God about your life. *Pray for strength to live your life the way Christ lived His!*

Read
-Devote four minutes to reading 1st Corinthians 11.

Reflect
-Take three minutes to think through the following questions.
1. Do you agree with the following quote? *"Example is not the best way to teach—it is the only way!"*
2. Read verse 1—Can you make this statement to your family and friends?
3. If your family and friends followed in your footsteps, would they end up in heaven?
4. What is the purpose of the Lord's Supper? How should one partake of the Lord's Supper?

Record
-Spend three minutes answering the following questions.
1. Write down one question or observation you have over this chapter. _____
2. List one way this chapter will strengthen your daily relationship with Jesus. _____
3. Write out one action step you will take in your life based on today's reading. *I will* _____

Request
-Devote four minutes to talking with God.

Lesson 174

Reflect
-Spend one minute meditating on the forgiveness available through Jesus. *"Father forgive them for they do not know what they are doing." Luke 24:34*

Read
-Devote four minutes to reading Psalm 120 – Psalm 124.

Reflect
-Take three minutes to think through the following questions.
1. What is the most important lesson you learned from these Psalms?
2. Read Psalm 123:1-2—How can you keep your eyes focused on God in your daily life?
3. Is it difficult to always keep our eyes on God? Why or why not?
4. What is the importance of having God on our side in our daily life?

Record
-Spend three minutes answering the following questions.
1. Write down one question or observation you have over these Psalms. _____

2. List one way these Psalms will help you seek a more intimate relationship with God. _____

3. Write out one action step you will take in your life based on today's reading. *I will* _____

Request
-Devote four minutes to talking with God.

Lesson 175

Relax
-Spend one minute meditating on the following words.
"Travel the path of integrity without looking back, for there is never a wrong time to do the right thing."

Read
-Devote four minutes to reading 1st Corinthians 12.

Reflect
-Take three minutes to think through the following questions.
1. What analogy does Paul use to describe the importance of each individual Christian?
2. Do you agree with the following statement. *Every Christian is a minister—Every task is important—Every member is talented in some area!*
3. Are you currently using your talents/abilities to help the Lord's church grow?

Record
-Spend three minutes answering the following questions.
1. Write out one question or observation you have over this chapter. _____
_____
2. List two abilities God has given you, and how you plan to use these gifts to help God's kingdom grow.
_____
_____
3. Write out one action step you will take in your life based on today's reading. *I will* _____
_____

Request
-Devote four minutes to talking with God.

Lesson 176

Relax
-Spend one minute meditating on the following words.
"Jesus wants you to have the richest, fullest life you could ever imagine." John 10:10

Read
-Devote four minutes to reading Psalm 125 – Psalm 129.

Reflect
-Take three minutes to think through the following questions.
1. What is your definition of a rich, full life? Is your definition the same as God's?
2. Read Psalm 127:1—How can you allow God to build your house?
3. What is the connection between fearing God and being blessed by God?
4. How can one tell if they truly fear God? Why is it important that you fear God in your life?

Record
-Spend three minutes answering the following questions.
1. Write out one question or observation you have over these Psalms. _____

2. List one lesson you learned today that will help you fulfill your responsibilities to your family. _____

3. Write out one action step you will take in your life based on today's reading. *I will* _____

Request
-Devote four minutes to talking with God.

Lesson 177

Relax
-Devote one minute to meditating on the following quote. *You have just one life to live for God! Seize today, and make the most of it!*

Read
-Spend four minutes reading 1st Corinthians 13. *Try to memorize verses 4 – 8a.*

Reflect
-Take three minutes to think through the following questions.
1. How is God's definition of love different than the world's definition?
2. Read verses 4-7 again—In which of these areas are you strong? Weak?
3. What is the difference between loving someone and lusting after someone?
4. Read verse 13—Why does Paul say the gift of love is the greatest gift of all?

Record
-Spend three minutes answering the following questions.
1. Write down one question or observation you have over this chapter. _____

2. List one way this chapter will help you become a more effective minister. _____

3. Write out one action step you will take in your life based on today's reading. *I will* _____

Request
-Devote four minutes to talking with God.

Lesson 178

Relax

-Spend one minute meditating on the words of 1st Corinthians 13:8a while evaluating your relationship with God and your family. *"Love never fails!"*

Read

-Devote four minutes to reading Psalm 130 – Psalm 134.

Reflect

-Take three minutes to think through the following questions.
1. Read Psalm 130:3—What do we learn about the forgiveness of God in this verse?
2. How do you feel knowing God not only forgives our sins, but He also *forgets* them?
3. How can someone wait for the Lord? Hope in the Lord? Bless the Lord?
4. How does humility/lack of humility impact one's relationship with God?

Record

-Spend three minutes answering the following questions.
1. Write down one question or observation you have over these Psalms. _____
_____
2. List one lesson you learned today that will deepen your commitment to live for God. _____
_____
3. Write out one action step you will take in your life based on today's reading. *I will* _____
_____

Request

-Devote four minutes to talking with God.

Lesson 179

Relax
-Spend one minute meditating on the following words. *The formula for peace of mind is simple, worry less and pray more! Philippians 4:6, 7.*

Read
-Devote four minutes to reading 1st Corinthians 14.

Reflect
-Take three minutes to think through the following questions.
1. Why is there so much confusion in the religious world today?
2. What will it take to bring about unity from all of this confusion?
3. Read verse 20—Would you describe the Corinthians as being spiritually mature? What about you?
4. What was the purpose of spiritual gifts as described by Paul in this chapter?

Record
-Spend three minutes answering the following questions.
1. Write down one question or observation you have over this chapter. _____
2. List one lesson you learned today that will help you mature in your relationship with Jesus. _____
3. Write out one action step you will take in your life based on today's reading. *I will* _____

Request
-Devote four minutes to talking with God.

Lesson 180

Relax
-Spend one minute meditating on the significance of these words found in Psalm 135 and Psalm 136. *"The name of the Lord and the love of the Lord endures forever."*

Read
-Devote four minutes to reading Psalm 135 – Psalm 136.

Reflect
-Take three minutes to think through the following questions.
1. How many times did you read the phrase *"the love/mercy of the Lord endures forever"* in these Psalms?
2. Why is it important for you to remember these words?
3. How has God demonstrated the eternalness of His love to you in your life?
4. What are some things that can cause us to forget about God's love?

Record
-Spend three minutes answering the following questions.
1. Write out one question or observation you have over these Psalms. _____
2. List one lesson you learned today that will strengthen your love for God. _____
3. Write out one action step you will take in your life based on today's reading. *I will* _____

Request
-Devote four minutes to talking with God.

Lesson 181

Relax
-Spend one minute in a quiet place reflecting on one positive change Jesus has brought into your life.

Read
-Devote four minutes to reading 1st Corinthians 15.

Reflect
-Take three minutes to think through the following questions.
1. Where should we take our stand as Christians?
2. What is the gospel? How would you explain the power of the gospel to a friend or relative?
3. What is the significance of Jesus' resurrection to Christians?
4. What are some things Jesus provides victory over for you as a Christian?
5. Read verse 31—What must you do in your life to be able to write the words Paul wrote in this verse?

Record
-Spend three minutes answering the following questions.
1. Write down one question or observation you have over this chapter. _____
   _____

2. List one lesson you learned from this chapter that will die to self and live for Jesus. _____
   _____

3. Write out one action step you will take in your life based on today's reading. *I will* _____
   _____

Request
-Devote four minutes to talking with God.

Lesson 182

Relax
-Spend one minute meditating on how you feel knowing you will never be alone because of God's continual presence in your life.

Read
-Devote four minutes to reading Psalm 137 – Psalm 139.

Reflect
-Take three minutes to think through the following questions.
1. Read Psalm 138:7—What is the significance of this verse to your life?
2. How well does God know you individually?
3. What did you learn about the presence of God in Psalm 139?
4. Have you ever wanted to escape from the presence of God?
5. Is it possible to reject God's presence in your life?

Record
-Spend three minutes answering the following questions.
1. Write down one question or observation you have over these Psalms. _____

2. List one lesson you learned about God that will strengthen your desire to live for Him. _____

3. Write out one action step you will take in your life based on today's reading. *I will* _____

Request
-Devote four minutes to talking with God.

Lesson 183

Relax
    -Spend one minute meditating on the words of Jesus found in John 14:27. *"Peace I leave with you; My peace I give to you...do not let your heart be troubled, nor let it be fearful."*

Read
    -Devote four minutes to reading 1st Corinthians 16.

Reflect
    -Take three minutes to think through the following questions.
1. What were the most powerful words Paul used in concluding this letter to the Corinthians?
2. Read verse 13—How can you as a Christian fulfill these words in your relationships with your family and God?
3. In verse 9, Paul talks about a ministry opportunity opened to him. *Are you praying for, and looking for open doors to walk through for God?*

Record
    -Spend three minutes answering the following questions.
1. Write down one question or observation you have over this chapter. _____
2. What is the most important lesson you learned from 1st Corinthians? _____
3. Write out one action step you will take in your life based on today's reading. *I will* _____

Request
    -Devote four minutes to talking with God.

Lesson 184

Relax
    -Spend one minute meditating on the following quote after reading James 1:2-4. *Out of struggle comes strength of character.*

Read
    -Devote four minutes to reading Psalm 140 – Psalm 142.

Reflect
    -Take three minutes to think through the following questions.
1. What struggles was David experiencing in Psalm 140? Have you ever had similar problems?
2. Read Psalm 142:4—Why was David's relationship with God so important to him?
3. What is the connection between verses 3 and 4 in Psalm 141 and verse 8 in Psalm 141?
4. When is it most difficult for you to keep your eyes fixed on God?

Record
    -Spend three minutes answering the following questions.
1. Write down one question or observation you have over these Psalms. _____
2. List one way these Psalms will help you keep your eyes fixed on God in your daily life. _____
3. Write out one action step you will take in your life based on today's reading. *I will* _____

Request
    -Devote four minutes to talking with God.

Lesson 185

Relax
-Spend one minute meditating on the following quote while reflecting on what matters most in your life. *There can be only one highest priority in life!*

Read
-Devote four minutes to reading Galatians 1.

Reflect
-Take three minutes to think through the following questions.
1. What are some of the causes of religious confusion today?
2. Read verse 10—Are you more concerned with pleasing humans (friends, spouse, parents) or pleasing God?
3. Have you made a mistake in your life that you've struggled to overcome?
4. What can we learn from Paul about overcoming past failures and mistakes?

Record
-Spend three minutes answering the following questions.
1. Write down one question or observation you have over this chapter._____
2. Which verse impacted you the most from today's reading? _____
3. Write out one action step you will take in your life based on today's reading. *I will* _____

Request
-Devote four minutes to talking with God.

Lesson 186

Relax
    -Spend one minute meditating on the significance of Paul's words in Galatians 2:20 in relationship to developing a powerful daily walk with God.

Read
    -Devote four minutes to reading Galatians 2.

Reflect
    -Take three minutes to think through the following questions.
1. Which verse(s) from this chapter indicates that Paul was experiencing some doubts in his ministry?
2. Is it sinful to have doubts in our ministry? How do you handle doubts when they arise?
3. Read Galatians 2:20—Have you completely died to self and committed your life completely to God?
4. What is one area of your life that you struggle to give up to God?

Record
    -Spend three minutes answering the following questions.
1. Write down one question or observation you have over this chapter. _____

2. List one way this chapter will strengthen your daily walk with Jesus. _____

3. Write out one action step you will take in your life based on today's reading. *I will* _____

Request
    -Devote four minutes to talking with God.

Lesson 187

Relax
-Spend one minute meditating on Jeremiah 20:9b. *"His Word is in my heart like a fire, a fire shut up in my bones. I am weary of holding it in; indeed I cannot."*

Read
-Devote four minutes to reading Galatians 3.

Reflect
-Take three minutes to think through the following questions.
1. What are some blessings you have only through Jesus? What curse has Jesus freed you from?
2. Read verse 24—What is the purpose of the Old Testament for Christians today?
3. Why is Abraham referred to as "the man of faith" in verse 9 of this chapter?
4. Would you describe yourself as a man or woman of faith? Would God describe you in this manner?

Record
-Spend three minutes answering the following questions.
1. Write down one question or observation you have over this chapter. _____

2. List one way this chapter will increase your gratitude for Jesus. _____

3. Write out one action step you will take in your life based on today's reading. *I will* _____

Request
-Devote four minutes to talking with God.

Lesson 188

Relax
    -Spend one minute meditating on the words of David in Psalm 144:15b. *"Blessed are the people whose God is the Lord."*

Read
    -Devote four minutes to reading Psalm 143 – Psalm 144.

Reflect
    -Take three minutes to think through the following questions.
        1. Reflect on a time in your life when you were in desperate need of mercy. *Is it easier for you to ask for mercy or extend mercy? Why?*
        2. We all need God's mercy! *What did you learn about God's nature in Psalm 143 that should give you confidence in asking God for mercy?*
        3. Read Psalm 143:10—*Who or what is your god? Can you honestly say the Lord is your God?*

Record
    -Spend three minutes answering the following questions.
        1. Write down one question or observation you have over these Psalms. _____

        2. Read Psalm 144:4—How does this verse impact your desire to live for God? _____

        3. Write out one action step you will take in your life based on today's reading. *I will* _____

Request
    -Devote four minutes to talking with God.

Lesson 189

Relax
    -Spend one minute in a quiet place talking with God about your life. *Share with Him any problems or concerns you have in your relationships with your family and friends.*

Read
    -Devote four minutes to reading Galatians 4.

Reflect
    -Take three minutes to think through the following questions.
      1. Read verse 9—Based on the actions of the Galatians, did you think they knew God? Why or why not?
      2. God knows everything about you. *How deeply do you know God? How can we tell if we truly know God?*
      3. The Galatians were turning back to their old ways of following the Law. *Do Christians today face similar temptations? Explain.*

Record
    -Spend three minutes answering the following questions.
      1. Write down one question or observation you have over this chapter. _____

      2. Which verse from today's reading had the biggest impact on your life? _____

      3. Write out one action step you will take in your life based on today's reading. *I will* _____

Request
    -Devote four minutes to talking with God.

Lesson 190

Relax
- Spend one minute meditating on the words of Jesus in Matthew 7:8. *"For everyone who asks receives, and he who seeks finds, and to him who knocks it shall be opened."*

Read
- Devote four minutes to reading Galatians 5.

Reflect
- Take three minutes to think through the following questions.
    1. What are some things that can knock us out of the Christian race?
    2. *"Love your neighbor as yourself."* What impact do these words have on your outreach to the lost?
    3. Which of the sins of the flesh listed in verses 19 – 21 do you struggle with the most in your life?
    4. Which verse in this chapter is the key to overcoming our sinful desires?

Record
- Spend three minutes answering the following questions.
    1. Write down one question or observation you have over this chapter. _____
    2. Which characteristic of the fruit of the Spirit is most abundant in your life? Which one is lacking? _____
    3. Write down one action step you will take in your life based on today's reading. *I will* _____

Request
- Devote four minutes to talking with God.

Lesson 191

Relax
-Spend one minute meditating on the words of Galatians 6:9. *"Let us not become weary in doing good, for at the proper time we will reap a harvest if we do not give up."*

Read
-Devote four minutes to reading Galatians 6.

Reflect
-Take three minutes to think through the following questions.
1. Read Galatians 2:20; 5:24; and 6:14—What do these verses have in common?
2. What are some things you need to crucify in your life?
3. What responsibilities do Christians have to others?
4. Read verse 2 and verse 5—Why does Paul write these two seemingly contradictory statements?

Record
-Spend three minutes answering the following questions.
1. Write down one question or observation you have over this chapter. _____

2. List one way this chapter will help you overcome the daily temptations you face. _____

3. Write out one action step you will take in your life based on today's reading. *I will* _____

Request
-Devote four minutes to talking with God asking specifically for strength in dying to your selfish desires.

Lesson 192

Relax
-Spend one minute meditating on the words of Psalm 145:3. *"Great is the Lord and most worthy of praise; His greatness no one can fathom."*

Read
-Devote four minutes to reading Psalm 145.

Reflect
-Take three minutes to think through the following questions.
1. What did you learn about God's character from this Psalm?
2. Is there a character quality you've placed on God that is perhaps untrue or unfair?
3. When did you begin to grasp a clear understanding of the depth of God's true character?
4. What is the significance of verse 18 in your relationship with God?

Record
-Spend three minutes answering the following questions.
1. Write down one question or observation you have over this Psalm. _____

2. List one way this Psalm will help you strengthen your relationship with God. _____

3. Write out one action step you will take in your life based on today's reading. *I will* _____

Request
-Devote four minutes to talking with God.

Lesson 193

Relax
-Spend one minute meditating on the love of Christ. *"In Him we have redemption through His blood, the forgiveness of sins, according to the riches of His grace."* Ephesians 1:7

Read
-Devote four minutes to reading Ephesians 1.

Reflect
-Take three minutes to think through the following questions.
1. What are some benefits that only come through a relationship with Jesus?
2. How do we enter a saving relationship with Jesus? (Galatians 3:26-28; Romans 6:1-7)
3. What is your greatest source of motivation to live for Jesus?
4. Read verse 11—In what ways does this verse motivate you as a Christian?

Record
-Spend three minutes answering the following questions.
1. Write down one question or observation you have over Ephesians 1. _____

2. Describe the relationship between Jesus and the church. _____

3. Write out one action step you will take in your life based on today's reading. *I will* _____

Request
-Devote four minutes to talking with God.

Lesson 194

Relax

-Spend one minute meditating on the words of David in Psalm 145:8. *"The Lord is gracious and full of compassion, slow to anger and great in mercy."*

Read

-Devote four minutes to reading Ephesians 2.

Reflect

-Take three minutes to think through the following questions.
1. How would you describe God's grace based on what you read in this chapter?
2. What is the relationship between your faith and your actions/works?
3. Read verse 10—Why has God created you? Are you currently fulfilling God's purpose for your life?
4. Reflect on the words of verse 22. *"You are being built together for a dwelling place of God..."*

Record

-Spend three minutes answering the following questions.
1. Write down one question or observation you have over this chapter. _____
_____
2. Which verse from Ephesians 2 had the greatest impact on you? Why? _____
_____
3. Write out one action step you will take in your life based on today's reading. *I will* _____
_____

Request

-Devote four minutes to talking with God.

Lesson 195

Relax
    -Spend one minutes meditating on the following words.
    "There is no road to success but through a clear, strong purpose."

Read
    -Devote four minutes to reading Ephesians 3.

Reflect
    -Take three minutes to think through the following questions.
        1. How would you describe the purpose of Jesus' life? What about the apostle Paul? What about your life?
        2. What are some obstacles that have prevented you from accomplishing God's purpose for your life?
        3. Read verse 17—Have you made room in your heart for Jesus?
        4. What is the significance of verse 20 to you personally?

Record
    -Spend three minutes answering the following questions.
        1. Write down one question or observation you have over this chapter. _____
        2. List one way Ephesians 2 will strengthen your desire to follow Christ. _____
        3. Write out one action step you will take in your life based on today's reading. *I will* _____

Request
    -Devote four minutes to talking with God.

Lesson 196

Relax
-Spend one minute meditating on the words of Psalm 18:2.
"The Lord is my rock, my fortress and my deliverer."

Read
-Devote four minutes to reading Psalm 146 and Psalm 147.

Reflect
-Take three minutes to think through the following questions.
1. Have you ever considered that you have the ability to bring God pleasure?
2. Read Psalm 147:11—Are you currently bringing pleasure to God?
3. Read Psalm 147:5—How does the Psalmist describe God's understanding?
4. How do you feel knowing you have direct access to an all knowing, all understanding Father?

Record
-Spend three minutes answering the following questions.
1. Write down one question or observation you have over these Psalms. _____

2. How would you describe God's power? _____

3. List one way these Psalms have increased your knowledge of who God is. _____

4. Write out one action step you will take in your life based on today's reading. *I will* _____

Request
-Devote four minutes to talking with God.

Lesson 197

Relax
-Spend one minute meditating on the power of God. *David did not look at the size of the obstacle, but at the power of his God.*

Read
-Devote four minutes to reading Ephesians 4.

Reflect
-Take three minutes to think through the following questions.
1. How important is unity to God based on what you read in this chapter?
2. What is the key to achieving spiritual unity in the religious world today?
3. What did you learn from this chapter about the importance of spiritual growth?
4. Read verse 29—Are you fulfilling this verse in your life?

Record
-Spend three minutes answering the following questions.
1. Write down one question or observation you have over this chapter. _____
2. Which verse from Ephesians 4 is most challenging for you to follow in your daily walk? _____
3. Write out one action step you will take in your life based on today's reading. *I will* _____

Request
-Devote four minutes to talking with God.

Lesson 198

Relax
 -Spend one minute visualizing the following words in your mind. *"There could be no greater reward than seeing Jesus smile at the events of your life when all is said and done!"*

Record
 -Devote four minutes to reading Ephesians 5.

Reflect
 -Take three minutes to think through the following questions.
 1. Is it possible to imitate God without first knowing Him? Are you able to imitate God?
 2. Read verse 3—Which of these sins do you struggle the most with in your life—immorality, impurity or greed?
 3. After reading this chapter, can you honestly say you are fulfilling your responsibilities to your spouse?
 4. What is one thing you can do to become a more godly spouse?

Record
 -Spend three minutes answering the following questions.
 1. Write down one question or observation you have over this chapter. _____
 2. Compare Jesus' love for the church with your love for the church. _____
 3. Write out one action step you will take in your life based on today's reading. *I will* _____

Request
 -Devote four minutes to talking with God.

Lesson 199

Relax
-Spend one minute in a quiet place talking with your Father about one temptation you are currently struggling to overcome.

Read
-Devote four minutes to reading Ephesians 6.

Reflect
-Take three minutes to think through the following questions.
1. How would you describe Satan? Is it difficult for you to fight against Satan?
2. What do each of the pieces of God's armor have in common?
3. Which piece of God's armor is most difficult for you to put on in your life?
4. What is the role of prayer in our spiritual battle against Satan? How did Jesus use prayer to prepare Himself for the battles He faced? (Matthew 26)

Record
-Spend three minutes answering the following questions.
1. Write down one question or observation you have over this chapter. _____
   _____
2. How successful are you at overcoming Satan's attacks? _____
3. Write out one action step you will take in your life based on today's reading. *I will* _____
   _____

Request
-Devote four minutes to talking with God.

Lesson 200

Relax
-Spend one minute thinking of reasons why God is deserving of your praise. *"Let everything that has breath praise the Lord."*

Read
-Devote four minutes to reading Psalm 148 – Psalm 150.

Reflect
-Take three minutes to think through the following questions.
1. Is God being praised based on how you are currently living your life?
2. Who would God rather receive praise from; the mountains or you?
3. Reflect on the following statement: Of all the wondrous things God has created, you are #1!
4. In what ways have you been blessed from reading and studying the Psalms?

Record
-Spend three minutes answering the following questions.
1. Write down one question or observation you have over these Psalms. _____
2. List one way these Psalms will strengthen your relationship with God. _____
3. Write out one action step you will take in your life based on today's reading. *I will* _____

Request
-Devote four minutes to talking with God.

Lesson 201

Relax
    -Spend one minute meditating on the significance of the words found in Luke 1:37 for your life. *"For nothing will be impossible with God."*

Read
    -Devote four minutes to reading Luke 1:1-45.

Reflect
    -Take three minutes to think through the following questions.
1. What are some examples from these verses of God's power in action?
2. Do you believe the words written in Luke 1:37? *Imagine what you can accomplish in your life through God!*
3. What would you attempt for Jesus if you knew you could never fail?
4. Is your faith more similar to Mary's or Zacharias'?

Record
    -Spend three minutes answering the following questions.
1. Write down one question or observation you have over these verses. _____

2. List one way today's reading will help you grow in your faith. _____

3. Write out one action step you will take in your life based on today's reading. *I will* _____

Request
    -Devote four minutes to talking with God.

Lesson 202

Relax
    -Spend one minute thinking on the following quote in relation to a struggle you are currently facing. *If God brings you to it--He will bring you through it!* 1st Corinthians 10:13

Read
    -Devote four minutes to reading Luke 1:46-80.

Reflect
    -Take three minutes to think through the following questions.
1. What were some miracles associated with the birth John?
2. How would you describe the mission of John?
3. What are some things you want to accomplish in your life?
4. What are some goals you want to accomplish for Jesus? *Are you living your life to achieve these goals?*

Record
    -Spend three minutes answering the following questions.
1. Write down one question or observation you have over these verses. _____
2. Write out your personal mission statement as a Christian. _____
3. List one action step you will take in your life based on your reading from Luke 1. *I will* _____

Request
    -Devote four minutes to talking with God.

Lesson 203

Relax
-Spend one minute meditating on the following words from Proverbs. *"A wise man will hear and increase in learning, and a man of understanding will acquire wise counsel."*

Read
-Devote four minutes to reading Proverbs 1.

Reflect
-Take three minutes to think through the following questions.
1. Who wrote the Proverbs? What are some secrets to gaining wisdom according to this first proverb?
2. Read 1st Kings 3:6-12—How would you describe the wisdom of Solomon? Where did Solomon gain his wisdom?
3. What is the greatest piece of advice you discovered from this proverb?

Record
-Spend three minutes answering the following questions.
1. Write down one question or observation you have over Proverbs 1. _____

2. List one way this proverb will help you become more like Christ in your decision making. _____

3. Write out one action step you will take in your life based on today's reading. *I will* _____

Request
-Devote four minutes to talking with God asking specifically for a wise and discerning heart as Solomon did.

Lesson 204

Relax
-Spend one minute meditating on the following words. *"You don't invent your talents in life, you simply detect them."*

Read
-Devote four minutes to reading Luke 2.

Reflect
-Take three minutes to think through the following questions.
1. What is the most interesting lesson you learned from Luke 2 about Jesus' early life as a child?
2. How important were obeying the commands of God to Mary and Joseph? *How important was obeying God to Jesus?*
3. What did you learn about Jesus from Simeon's statement about Him in verses 29-32?
4. How do you think Mary felt about the words said to her about her Son in verse 35?

Record
-Spend three minutes answering the following questions.
1. Write down one question or observation you have over this chapter. _____
2. List one way this chapter will deepen you love for God. _____
3. Write out one action step you will take in your life based on today's reading. *I will* _____

Request
-Devote four minutes to talking with God.

Lesson 205

Relax
-Spend one minute meditating on the words said to Jesus by His Father in Luke 3:22. *"You are my beloved Son in You I am well pleased."*

Read
-Devote four minutes to reading Luke 3.

Reflect
-Take three minutes to think through the following questions.
1. Are you living your life in such a way that God can say to you what He said to Jesus in verse 22?
2. What message did John preach?
3. How do we know if one has truly repented of their sins? (Read verse 8)
4. Was John a humble man? What did you discover in this chapter that reveals the humility of John? *Hint: read verses 15-17!*

Record
-Spend three minutes answering the following questions.
1. Write down one question or observation you have over Luke 3. _____

2. List one way this chapter will deepen your desire to live for Jesus. _____

3. Write out one action step you will take in your life based on today's reading. *I will* _____

Request
-Devote four minutes to talking with God.

Lesson 206

Relax
-Spend one minute meditating on the following words. *If you seek wisdom, understanding, and discernment from God— you will find it!*

Read
-Devote four minutes to reading Proverbs 2.

Reflect
-Take three minutes to think through the following questions.
1. What are some things people are searching for today?
2. Are people today more apt to search for physical treasure or spiritual treasure? *What are you searching for in your life?*
3. What is the key to overcoming temptation according to Solomon in this proverb?
4. Read verse 3—Have you ever done these things?

Record
-Spend three minutes answering the following questions.
1. Write down one question or observation you have over this proverb. _____
2. Explain the connection between seeking God and avoiding sin. _____
3. Write out one action step you will take in your life based on today's reading. *I will* _____

Request
-Devote four minutes to talking with God.

Lesson 207

Relax
-Spend one minute meditating on the words of Luke 4:13 in relation to your daily battles against Satan. *"When the devil had finished every temptation, he left Him……."*

Read
-Devote four minutes to reading Luke 4.

Reflect
-Take three minutes to think through the following questions.
1. How would you define the word 'temptation?' Is it sinful to be tempted?
2. In your opinion which one of these temptations was most 'tempting' to Jesus?
3. What is the biggest temptation you are currently facing?
4. What did you learn about Satan in this chapter that will help you overcome temptations?
5. How would you describe Jesus' ministry based on what you read in this chapter?

Record
-Spend three minutes answering the following questions.
1. Write down one question or observation you have over this chapter. _____
   _____
2. Why did the people in the synagogue try to throw Jesus off a cliff? _____
   _____
3. Write out one action step this chapter has motivated you to take in your life. *I will* _____
   _____

Request
-Devote four minutes to talking with God.

Lesson 208

Relax
-Spend one minute meditating on the significance of the words spoken by Jesus to the paralytic in Luke 5:20. *"Friend, your sins are forgiven you."*

Read
-Devote four minutes to reading Luke 5.

Reflect
-Take three minutes to think through the following questions.
1. What motivated Peter to drop everything and follow Jesus? What about Matthew?
2. Read verse 16—How would you describe the role of prayer in the life of Jesus?
3. Do you take time in your life to 'slip away' and talk with your Father?
4. What is the answer to the question Jesus asked the Pharisees in verse 23?

Record
-Spend three minutes answering the following questions.
1. Write down one question or observation you have over this chapter. _____

2. What is the most powerful lesson you learned from studying this chapter? _____

3. Write out one action step this chapter has motivated you to take in your life. *I will* _____

Request
-Devote four minutes to talking with God.

Lesson 209

Relax
    -Spend one minute meditating on the words of Proverbs 3:5. *"Trust in the Lord with all your heart and do not lean on your own understanding."*

Read
    -Devote four minutes to reading Proverbs 3.

Reflect
    -Take three minutes to think through the following questions.
1. How can we accomplish the words Solomon wrote in verses 5 and 6?
2. What is the connection between knowing God and trusting God?
3. Do you currently have a close, intimate relationship with God?
4. Read verse 32—What must you do in order for God to have an intimate relationship with you?

Record
    -Spend three minutes answering the following questions.
1. Write down one question or observation you have over this proverb. _____

2. Which teaching from this proverb is most relevant, and meaningful to your life? _____

3. Write out one action step you will take in your life based on today's reading. *I will* _____

Request
    -Devote four minutes to talking with God.

Lesson 210

Relax
-Spend one minute in a quiet place talking with God about your spiritual walk. *Share any concerns or struggles you are facing with your Father.*

Read
-Devote four minutes to reading Luke 6.

Reflect
-Take three minutes to think through the following questions.
1. Why did Jesus spend the entire night in prayer? *What important decision was He preparing to make?*
2. Are you currently using prayer to prepare you to make important decisions in your life?
3. What is the difference between the two men Jesus describes in verses 46-49?
4. Which one of these men best describes your current spiritual state? *What are you doing for Jesus?*

Record
-Spend three minutes answering the following questions.
1. Write down one question or observation you have over Luke 6. _____

2. Which one of Jesus' teaching in this chapter is most challenging for you to follow? _____

3. Write out one action step this chapter has motivated you to take in your life. *I will* _____

Request
-Devote four minutes to talking with God.

Lesson 211

Relax
-Spend one minute meditating on the following words from Luke. *"The blind receive sight, the lame walk, the lepers are cleansed, and the deaf hear....."* **Jesus Heals!**

Read
-Devote four minutes to reading Luke 7.

Reflect
-Take three minutes to think through the following questions.
1. Why was Jesus amazed at the faith at the centurion?
2. How would you describe your faith? *Do you think your faith has ever amazed Jesus?*
3. Read verse 16—Does this community see us the way the city of Nain saw Jesus? Why or why not?
4. Read verse 30—Have you ever been guilty of rejecting God's purpose for your life?
5. What is God's most important purpose for your life?

Record
-Spend three minutes answering the following questions.
1. Write down one question or observation you have over this chapter. _____

2. List one way this chapter has strengthened your love for Jesus. _____

3. Write out one action step you will take in your life based on this chapter. *I will* _____

Request
-Devote four minutes to talking with God.

Lesson 212

Relax
-Spend one minute meditating on the words of Proverbs 4:23. *"Watch over your heart with all diligence, for from it flow the springs of life."*

Read
-Devote four minutes to reading Proverbs 4.

Reflect
-Take three minutes to think through the following questions.
1. Why is it crucial for Christian's to guard their hearts against evil?
2. How can we guard our hearts against evil?
3. Solomon tells us how we can guard our hearts in verses 24-27. *Which one of these teachings is most challenging for you to accomplish in your life?*
4. What is the connection between your speech, your thoughts, your actions and your heart?

Record
-Spend three minutes answering the following questions.
1. Write down one question or observation you have over this proverb. _____

2. List the most important lesson you learned from this Proverbs 4. _____

3. Write out one action step you will take in your life as a result of today's study. *I will* _____

Request
-Devote four minutes to talking with God.

Lesson 213

Relax
  -Spend one minute meditating on the following quote.
  Everything can be taken from a man but one thing—the last of the human freedoms—to choose one's attitude in any given set of circumstances.

Read
  -Devote four minutes to reading Luke 8.

Reflect
  -Take three minutes to think through the following questions.
  1. How powerful is Jesus Christ?
  2. What is the most important lesson you learned from Jesus' parable of the sower in verses 4-15?
  3. How would you describe the impact Jesus made on the lives He came into contact with in this chapter?
  4. Read verse 39—What are some great things God has done for you? *Are you sharing these things with others?*

Record
  -Spend three minutes answering the following questions.
  1. Write down one question or observation you have over this chapter. _____
  _____
  2. List one way this chapter has motivated you to continue following in the footsteps of Jesus. _____
  _____
  3. Write out one action step you will take in your life based on today's reading. *I will* _____
  _____

Request
  -Devote four minutes to talking with God.

Lesson 214

Relax
-Spend one minute meditating on the following quote.
"Friendship along with love make life worth living." You have both of these things in God!

Read
-Devote four minutes to reading Luke 9.

Reflect
-Take three minutes to think through the following questions.
1. What is the most important step in becoming a committed disciple of Jesus?
2. Read verse 23—What does Jesus describe as the key step in being able to follow in His footsteps?
3. Have you considered that the biggest obstacle you'll face in developing a powerful relationship with Jesus is yourself?
4. The key to denying self is dying to self! *Are you fulfilling the words of Jesus in verse 23?*

Record
-Spend three minutes answering the following questions.
1. Write down one question or observation you have over Luke 9. _____
2. What is the importance of talking about Jesus in our everyday lives? _____
3. Write down one action step this chapter has motivated you to take in your life. *I will* _____

Request
-Devote four minutes to talking with God.

Lesson 215

Relax
-Spend one minute in a quiet place talking with your Father. *Talk with God about your relationships with your family and You relationship with Him. Share your heart with God!*

Read
-Devote four minutes to reading Proverbs 5.

Reflect
-Take three minutes to think through the following questions.
1. What is the most important lesson you learned from reading this chapter?
2. How would you describe the nature of sin and temptation based on what you read in Proverbs 5?
3. How will this chapter improve your relationship with your spouse or help you get prepared to enter a marriage relationship?
4. What are God's expectations for the marriage relationship? Is society following God's plan?

Record
-Spend three minutes answering the following questions.
1. Write down one question or observation you have concerning Proverbs 5. _____
2. List one way this chapter will positively impact your daily walk with God. _____
3. Write out one action step this proverb has motivated you to take in your life. *I will* _____

Request
-Devote four minutes to talking with God.

Lesson 216

Relax
-Spend one minute meditating on the words Jesus said to His apostles in Luke 10:20. *"Rejoice that your names are recorded in heaven."*

Read
-Devote four minutes to reading Luke 10.

Reflect
-Take three minutes to think through the following questions.
1. How would you describe the mission Jesus gave to the seventy in verses 1-16?
2. Can Jesus look at you and how you are currently living and tell you to rejoice because your name is recorded in heaven?
3. What is the connection between gaining eternal life in heaven and verse 27?
4. Read verses 40-42—Is your life too busy to spend quality time with Jesus?

Record
-Spend three minutes answering the following questions.
1. Write down one question or observation you have over this chapter. _____

2. If you would have been one of the seventy, would you have accepted Jesus' mission? Explain. _____

3. Write out one action step you will take in your life based on today's reading. *I will* _____

Request
-Devote four minutes to talking with God.

Lesson 217

Relax
-Spend one minute meditating on the following quote. *"Life changes when you least expect it to. The future is uncertain. So, seize this day, seize this moment and make the most of it."*

Read
-Devote four minutes to reading Luke 11.

Reflect
-Take three minutes to think through the following questions.
1. What did you learn about prayer from Jesus' words in verses 2-4?
2. Is it easy or difficult for you to talk to God in prayer?
3. Read verse 34—Why does Jesus refer to the eye as the lamp of the body?
4. How would you describe the difference(s) between the Pharisees and Jesus?

Record
-Spend three minutes answering the following questions.
1. Write down one question or observation you have over this chapter. _____

2. List one way this chapter will deepen your love for Jesus. _____

3. Write out one action step you will take in your life based on today's reading. *I will* _____

Request
-Devote four minutes to talking with God in prayer.

Lesson 218

Relax
-Spend one minute meditating on the words of 1st Corinthians 3:11. *"For no man can lay a foundation other than the one which is laid, which is Jesus Christ."*

Read
-Devote four minutes to reading Proverbs 6.

Reflect
-Take three minutes to think through the following questions.
1. What did you learn about repairing relationships from this proverb?
2. Read verses 16-19—Which one of these things do you struggle with the most in your life?
3. How would you describe God's feelings towards sin?
4. What responsibilities do parents have to their children as described in this proverb?

Record
-Spend three minutes answering the following questions.
1. Write down one question or observation you have over this chapter. _____

2. What is the most important lesson you learned from Proverbs 6? _____

3. Write out one action step this proverb has motivated you to take in your life. *I will*_____

Request
-Devote four minutes to talking with God.

Lesson 219

Relax
-Spend one minute meditating on the words of Jesus from Luke 12:34. *"For where your treasure is, there your heart will be also."* Where's your heart?

Read
-Devote four minutes to reading Luke 12.

Reflect
-Take three minutes to think through the following questions.
1. What is one thing you are currently worrying about in your life? Why do people worry?
2. What did you learn about worry from this chapter?
3. What is a heavenly treasure? How can we store up treasures in heaven? (see verse 8)
4. Are you investing more time and energy storing up treasures in heaven or on earth?

Record
-Spend three minutes answering the following questions.
1. Write down one question or observation you have over this chapter. _____

2. List one way this chapter will help you in your daily walk with Jesus. _____

3. Write out one action step this chapter has motivated you to take in your life. *I will* _____

Request
-Devote four minutes to talking with God.

Lesson 220

Relax
-Spend one minute meditating on the condition of your spiritual heart while reading Matthew 5:8. *"Blessed are the pure in heart for they will see God."*

Read
-Devote four minutes to reading Luke 13.

Reflect
-Take three minutes to think through the following questions.
1. What is the importance of repentance in our relationship with God?
2. What does repentance involve?
3. Why is it sometimes difficult for people to repent and turn their lives over to God?
4. How is the opportunity to repent like a gift?
5. Why does Jesus compare the kingdom of God (the church) to a mustard seed?

Record
-Spend three minutes answering the following questions.
1. Write down one question or observation you have over this chapter. _____

2. List one way this chapter will strengthen your relationship with Jesus. _____

3. Write down one action step this chapter has motivated you to take in your life. *I will* _____

Request
-Devote four minutes to talking with God.

Lesson 221

Relax
    -Spend one minute meditating on the following quote. *"Your goals will determine where you go in life."* Take time to write down one personal and spiritual goal.

Read
    -Devote four minutes to reading Proverbs 7.

Reflect
    -Take three minutes to think through the following questions.
        1. How strong is your current relationship with God?
        2. What is the biggest obstacle preventing you from developing a stronger relationship with God?
        3. How was Jesus able to overcome all of the temptations He faced?
        4. Read Genesis 39:6-9—The key to overcoming temptation is to build a powerful relationship with _____.

Record
    -Spend three minutes answering the following questions.
        1. Write down one question or personal observation you have over this proverb. _____

        2. List one way this study will help you overcome daily temptations. _____

        3. Write down one action step you will take in your life based on today's reading. *I will* _____

Request
    -Devote four minutes to talking with God.

Lesson 222

Relax
-Spend one minute meditating on the following quote.
"Courage gives a Christian the ability to stand straight and not sway no matter which way the wind blows."

Read
-Devote four minutes to reading Luke 14.

Reflect
-Spend three minutes thinking on the following questions.
1. What are some situations where it takes courage to stand up for Christ?
2. Does it cost to be a Christian? What exactly does Christianity cost?
3. How would you convince a non-Christian that Christianity is worth the cost and sacrifice?
4. What is the greatest blessing you have as a Christian?

Record
-Take three minutes to think through the following questions.
1. Write down one question or personal observation you have over this chapter. _____

2. List one way this chapter will strengthen your desire to live for God. _____

3. Write out one action step this chapter has motivated you to take in your life. *I will* _____

Request
-Devote four minutes to talking with God.

Lesson 223

Relax
    -Spend one minute meditating on the words of Jesus in Luke 15:10. *"There is joy in the presence of the angels of God over one sinner who repents."*

Read
    -Devote four minutes to reading Luke 15.

Reflect
    -Take three minutes to think through the following questions.
1. How would you describe the attitude of the son upon leaving his father?
2. What caused this boy to realize the desperateness of his situation?
3. What feelings do you have for the father in this story?
4. Have you ever considered that your actions can cause the angels in heaven to rejoice?

Record
    -Spend three minutes answering the following questions.
1. Write down the most important lesson you learned from Luke 15. _____

2. List one way this chapter will deepen your love for God, your Father. _____

3. Write out one action step you will take in your life based on this chapter. *I will* _____

Request
    -Devote four minutes to talking with God.

Lesson 224

Relax
-Spend one minute meditating on the following quote.
"People have been known to achieve more as a result of working with others than against them."

Read
-Devote four minutes to reading Proverbs 8.

Reflect
-Take three minutes to think through the following questions.
1. What is the greatest team you have ever witnessed?
2. What is the greatest team you have ever been on?
3. What is the #1 key to building a successful team?
4. Have you ever considered that the greatest, powerful team you will ever be on is the Lord's team?
5. What are some lessons from Proverbs 8 that are essential in building a powerful team?

Record
-Spend three minutes answering the following questions.
1. Write down one question or observation you have over this chapter. _____
_____
2. What is the most important lesson you learned from Proverbs 8? _____
_____
3. Write out one action step you will take in your life based on today's reading. *I will* _____
_____

Request
-Devote four minutes to talking with God.

Lesson 225

Relax
    -Spend one minute meditating on heaven. *How awesome will it be to dwell in a place where there is no sorrow? "God will wipe away every tear from your eyes!"*

Read
    -Devote four minutes to reading Luke 16.

Reflect
    -Take three minutes to think through the following questions.
        1. Read verses 10-12: Can God trust you with the 'big' things?
        2. Have you ever tried to do something impossible? *If your trying to serve both God and money you are!*
        3. What did you learn about divorce and adultery from this chapter?
        4. How would you describe hell based on the experience of the rich man in this chapter?

Record
    -Spend three minutes answering the following questions.
        1. Write down one question or observation you have over Luke 16. _____

        2. List one way this chapter will help you in your personal ministry to others. _____

        3. Write out one action step you will take in your life based on today's reading. *I will* _____

Request
    -Devote four minutes to talking with God.

Lesson 226

Relax

-Spend one minute reflecting on two spiritual blessing you need to fall to your knees and thank God for. *"....and he (the leper) fell at his feet, giving thanks to Him."*

Read

-Devote four minutes to reading Luke 17.

Reflect

-Take three minutes to think through the following questions.
1. How would you describe the forgiveness of Jesus?
2. What did you learn about forgiveness from Luke 17?
3. Read verse 5: How can you increase your faith?
4. What do you have in common with the lepers?
5. Have you ever been guilty of taking Jesus for granted? *How often do you thank God for cleansing you of your sins?*

Record

-Spend three minutes answering the following questions.
1. Write down one question or observation you have over Luke 17. _____

2. List one way this chapter will strengthen your love for Jesus. _____

3. Write out one action step this chapter has motivated you to take in your life. *I will* _____

Request

-Devote four minutes to talking with God specifically thanking Him for *everything* He has given you!

Lesson 227

Relax
-Spend one minute meditating on the following quote. *"In the middle of difficulty lies opportunity."* James 1:2-3

Read
-Devote four minutes to reading Proverbs 9.

Reflect
-Take three minutes to think through the following questions.
1. What is the connection between humility and wisdom?
2. How does a wise man receive instruction/correction?
3. How do you typically react to someone when they offer you advice or instruction?
4. What impact does an arrogant attitude have on a relationship?
5. Do you agree with this statement: *Many people refuse to follow God because of their arrogance.*

Record
-Spend three minutes answering the following questions.
1. Write down one question or observation you have over this chapter. _____

2. List one key to gaining wisdom you learned from Proverbs 9. _____

3. Write out one action step you will take in your life based on today's reading. *I will* _____

Request
-Devote four minutes to talking with God.

Lesson 228

Relax
-Spend one minute meditating on the following words from Luke 18. *"The things that are impossible with people are possible with God."*

Read
-Devote four minutes to reading Luke 18.

Reflect
-Take three minutes to think through the following questions.
1. Read verse 8—If Jesus came today, would He find faith in your life? What about your family?
2. Is it easier to trust in self or trust in God?
3. How would you describe the man that talked with Jesus beginning in verse 18?
4. What is the most important lesson you learned from this man?
5. How can a person overcome selfishness?

Record
-Spend three minutes answering the following questions.
1. Write down one question or observation you have over this chapter. _____

2. List one thing you learned from this chapter that will help you deny yourself and follow Jesus. _____

3. Write out one action step you will take in your life based on today's reading. *I will* _____

Request
-Devote four minutes to talking with God.

Lesson 229

Relax
-Spend one minute meditating on the following words. *"Our greatest glory is not in never failing, but in rising up every time we fail."*

Read
-Devote four minutes to reading Luke 19.

Reflect
-Take three minutes to think through the following questions.
1. What are some things you have in common with Zaccheus?
2. How would you describe Jesus' purpose for coming to earth based on what you read in Luke 19?
3. Read verse 48—What impact do the words of Jesus have on your life?
4. What does Jesus expect you to do with the talents God has given you? *Are you using your talents?*

Record
-Spend three minutes answering the following questions.
1. Write down one question or observation you have over this chapter. _____

2. List the most important lesson you learned from Jesus' interaction with Zaccheus. _____

3. Write out one action step Luke 19 has motivated you to take in your life. *I will* _____

Request
-Devote four minutes to talking with your Father.

Lesson 230

Relax
-Spend one minute meditating on the following words. *"4th quarter, bottom of the 9th, last chance"—What would you do if this was your last day to live? Make it happen!*

Read
-Devote four minutes to reading Proverbs 10.

Reflect
-Take three minutes to think through the following questions.
1. Which contrast between the righteous and the wicked impacted you the most in this proverb?
2. How would you explain the differences between the heart of a wise man and the heart of a fool?
3. What attitude should Christians have towards material possessions? (Matthew 16:26)
4. What are some benefits of living a righteous life style?

Record
-Spend three minutes answering the following questions.
1. Write down one question or observation you have over Proverbs 10. _____

2. List one way this chapter will strengthen your daily relationship with God. _____

3. Write down one action step you will take in your life based on today's reading. *I will* _____

Request
-Devote four minutes to talking with God.

Lesson 231

Relax
-Spend one minute in a quiet place talking with your Father. *Talk with God about your spiritual growth. Share with Him one spiritual goal you are working towards achieving.*

Read
-Devote four minutes to reading Luke 20.

Reflect
-Take three minutes to think through the following questions.
1. Read verse 20—Have you ever been around someone who pretended to be righteous? Have you ever been guilty of doing this?
2. How does someone become righteous in the eyes of God?
3. Why did these religious leaders refuse to believe in Jesus as the Son of God?
4. Why do people reject Jesus today?

Record
-Spend three minutes answering the following questions.
1. Write down one question or observation you have over Luke 20. _____

2. List one way this chapter will help you strengthen your relationship with God. _____

3. Write out one action step this Bible study has motivated you to take in your life. *I will* _____

Request
-Devote four minutes to talking with God.

Lesson 232

Relax
   -Spend one minute meditating on your mission as a Christian while reading Philippians 2:14-15.

Read
   -Devote four minutes to reading Luke 21.

Reflect
   -Take three minutes to think through the following questions.
   1. What is the most important lesson you learned from this chapter?
   2. How would you describe the faith of the widow woman?
   3. What attitude did the people of this day (including the disciples) have toward the temple?
   4. What eventually happened to the temple and all of Jerusalem? How does this relate to us and our material possessions?

Record
   -Spend three minutes answering the following questions.
   1. Write down one question or observation you have over this chapter. _____

   2. List one way this chapter will strengthen your daily walk with Jesus. _____

   3. Write out one action step you will take in your life based on today's reading. *I will* _____

Request
   -Devote four minutes to talking with God.

Lesson 233

Relax
-Spend one minute meditating on the following quote.
"Faithfulness to God in the present allows Him to shape your heart for the future."

Read
-Devote four minutes to reading Proverbs 11.

Reflect
-Take three minutes to think through the following questions.
1. Have your prioritized your priorities? Is your life properly balanced with God in first place?
2. *Integrity is how we act when no one is looking!* Based on this definition, are you a person of integrity?
3. What are some lessons you learn about the tongue from this Proverb?
4. Read verse 27—What you seek you will find! *Are you seeking God or evil?*

Record
-Spend one minute answering the following questions.
1. Write down one question or observation you have over this Proverb. _____
2. List one way this chapter will help you become a more powerful minister for Jesus. _____
3. Write out one action step you will take in your life based on your study today. *I will* _____

Request
-Devote four minutes to talking with God.

Lesson 234

Relax
-Spend one minute meditating on the words of Proverbs 11:30 while *reflecting on the person or people who influenced you to become a Christian.*

Read
-Devote four minutes to reading Luke 22:1-38.

Reflect
-Take three minutes to think through the following questions.
1. How would you describe the work of Satan as seen in these verses?
2. What amazed you the most about Jesus in today's reading?
3. Why did Jesus institute the Lord's Supper with His apostles?
4. How does the world define greatness? How can one become great in the eyes of Jesus?

Record
-Spend three minutes answering the following questions.
1. Write down one question or observation you have over Luke 22:1-38. _____
_____
2. List one area in your spiritual life you need to strengthen. _____
_____
3. Write down one action step you will take in your life based on today's study. *I will* _____
_____

Request
-Devote four minutes to talking with your Father.

Lesson 235

Relax
-Spend one minute meditating on the words of Hebrews 10:23. *"Let us hold fast the confession of our hope without wavering, for He who promised is faithful..."*

Read
-Devote four minutes to reading Luke 22:39-71

Reflect
-Take three minutes to think through the following questions.
1. Why did Jesus encourage the disciples to pray?
2. What did Jesus use prayer to prepare Himself to face?
3. Are you currently using the prayer to prepare yourself for the battles you will face in life?
4. How would you describe Jesus' feeling towards His impending death on the cross?
5. Read verse 42—What is the significance of this verse for your life?

Record
-Spend three minutes answering the following questions.
1. Write down one question or observation you have over these verses. _____
2. What enabled Jesus to make the statement He made in verse 42? _____
3. Can you make this same statement? _____
4. Write out one action step these verse have motivated you to take in your life. *I will* _____
_____

Request
-Devote four minutes to talking with God.

Lesson 236

Relax
-Spend one minute meditating on the following words. *"If you have faith as small as a mustard seed—NOTHING will be impossible for you."*

Read
-Devote four minutes to reading Proverbs 12.

Reflect
-Take three minutes to think through the following questions.
1. What did you learn from this Proverb that will help you strengthen your relationships with your family and friends?
2. Read verse 5—How would you describe your thoughts? *What controls a person's thoughts?*
3. How can you change your thoughts? (Romans 12:1-2)
4. What does worry/anxiety do to a person's heart?

Record
-Spend three minutes answering the following questions.
1. Write down one question or observation you have over this Proverb. _____
2. List one way this chapter will help you become a more effective servant for God. _____
3. Write down one action step you will take in your life based on today's reading. *I will* _____

Request
-Devote four minutes to talking with God.

Lesson 237

Relax
-Spend one minute meditating on the words of Job 37:14.
"Stand still and consider the wondrous works of God."

Read
-Devote four minutes to reading Luke 23:1-32.

Reflect
-Take three minutes to think through the following questions.
1. Why did Pilate send Jesus to Herod?
2. How would you describe Herod based on what you read in these verses?
3. Did Pilate find Jesus guilty? What prevented Pilate from taking a stand for Jesus?
4. Are you willing to stand with Jesus at all costs?
5. What are some obstacles that have prevented you from standing up for Jesus in the past?

Record
-Spend three minutes answering the following questions.
1. Write down one question or observation you have over these verses. _____

2. List one lesson you learned from this reading that will strengthen your love for Jesus. _____

3. Write out one action step you will take in your life based on Luke 23:1-32. *I will* _____

Request
-Devote four minutes to talking with God *asking specifically for strength to stand up for Jesus at all cost.*

Lesson 238

Relax
    -Spend one minute meditating on the following quote. *"The mind is a powerful thing. How you think determines who you are!"*

Read
    -Devote four minutes to reading Luke 23:33-56.

Reflect
    -Take three minutes to think through the following questions.
        1. In your opinion, what was the greatest temptation Jesus faced in His life?
        2. What do you think Jesus was thinking as He was carrying His cross to His death?
        3. In what ways has the death of Jesus motivated you in your life?
        4. What powerful lesson do you learn about forgiveness from these verses?

Record
    -Spend three minutes answering the following questions.
        1. Write down one question or observation you have over these verses. _____

        2. What is the most important lesson you learn from Jesus' death on the cross? _____

        3. Write out one action step these verses have motivated you to take in your life. ***I will*** _____

Request
    -Devote four minutes to talking with your Father.

Lesson 239

Relax
    -Spend one minute meditating on people who are/were great Leaders while thinking on the following quote. *The most important place we can lead someone is to Christ!*

Read
    -Devote four minutes to reading Proverbs 13.

Reflect
    -Take three minutes to think through the following questions.
        1. What is the connection between love and discipline?
        2. Why is it important for parents to discipline their children?
        3. What are some different factors that motivate people to live for Christ?
        4. Read verse 13—In what ways does fear act as a motivator?
        5. Reflect on the following quote. *Love for God is the greatest motivator to live for God!*

Record
    -Spend three minutes answering the following questions.
        1. Write down one question or observation you have over this Proverb. _____

        2. List one way this reading will help you strengthen your daily relationship with God. _____

        3. Write out one action step this Proverb has motivated you to take in your life. *I will* _____

Request
    -Devote four minutes to talking with your Father.

Lesson 240

Relax
-Spend one minute meditating on the words of Colossians 3:16-17. *"All things were created through Him and for Him. He is before all things, and in Him all things hold together."*

Read
-Devote four minutes to reading Luke 24.

Reflect
-Take three minutes to think through the following questions.
1. Should the apostles and the other people have been surprised to see the tomb empty?
2. How would you describe the promises of Jesus?
3. Read verse 38—After all the experiences the apostles had with Jesus, why were they still battling doubt?
4. What are some situations that can bring about doubt and discouragement in a Christian's life?

Record
-Spend three minutes answering the following questions.
1. Write down one question or observation you have over Luke 24. _____
   _____

2. List one way this chapter has strengthened your faith in Jesus. _____
   _____

3. Write out one action step you will take in your life based on today's reading. *I will* _____
   _____

Request
-Devote four minutes to talking with God.

Lesson 241

Relax
-Spend one minute meditating on the words of Jeremiah 29:11. *"For I know the plans I have for you, declares the Lord,...plans to give you hope and a future."*

Read
-Devote four minutes to reading Proverbs 14.

Reflect
-Take three minutes to think through the following questions.
1. How would you define the phrase absolute truth?
2. Is there such a thing as absolute right and wrong?
3. Does society and religion in general today believe in absolute truth? Does God embrace absolute truth?
4. Read verse 12—Where can a person go to find the right way to live their life? (John 17:17)
5. Read verse 34—Would you describe our nation as sinful or righteous? Does your answer concern you?

Record
-Spend three minutes answering the following questions.
1. Write down one question or observation you have over this Proverb. _____
2. Which verse in your reading shows the importance of putting God's Word into action in our lives? ____
3. Write out one action step you will take in your life based on today's reading. *I will* _____

Request
-Devote four minutes to talking with God.

Lesson 242

Relax
-Spend one minute meditating on the words of Isaiah 40:31. *"Those who wait for the Lord will gain new strength; they will mount up with wings like eagles, they will run and not get tired, they will walk and not become weary."*

Read
-Devote four minutes to reading 2nd Corinthians 1.

Reflect
-Take three minutes to think through the following questions.
1. How many times is the word 'comfort' found in 2nd Corinthians 1? Why is this word used so often?
2. Why is it important for one to realize there is comfort in being a Christian?
3. What happened to Paul and others causing them to put their complete trust in God?
4. Take a moment to reflect on the words of verse 20.

Record
-Spend three minutes answering the following questions.
1. Write down one question or observation you have over this chapter. _____
2. List one way 2nd Corinthians 1 will strengthen your trust in God. _____
3. Write out one action step you will take in your life based on today's reading. *I will* _____

Request
-Devote four minutes to talking with God.

Lesson 243

Relax
    -Spend one minute meditating on a time in your life when God delivered you through a difficult struggle—*2nd Corinthians 1:10*.

Read
    -Devote four minutes to reading 2nd Corinthians 2.

Reflect
    -Take three minutes to think through the following questions.
1. What does Paul reveal about himself in this chapter that made him such an effective servant for God?
2. How did Paul feel about these Christians at Corinth?
3. Why is it important for a Christian to have a forgiving heart?
4. Would you describe yourself as a forgiving person?
5. Read verse 13—Are you willing to walk through the doors God opens for you in your life?

Record
    -Spend three minutes answering the following questions.
1. Write down one question or observation you have over 2nd Corinthians 2. _____
_____
2. List one way this chapter will help you prepare yourself for the attacks of Satan. _____
_____
3. Write out one action step you will take in your life based on today's reading. *I will* _____
_____

Request
    -Devote four minutes to talking with your Father.

Lesson 244

Relax
-Spend one minute meditating on the following words from Psalm 35. *"I lie down and sleep; I wake again because the Lord sustains me."*

Read
-Devote four minutes to reading Proverbs 15.

Reflect
-Take three minutes to think through the following questions.
1. Which verse(s) from today's reading impacted you the most?
2. What are some blessings found in this chapter for those who live a righteous life?
3. Read verse 3—Does this verse bring you confidence knowing God is with you, or does it bring you fear knowing God is watching you?
4. How important is humility in the Christian walk?

Record
-Spend three minutes answering the following questions.
1. Write down one question or observation you have over Proverbs 15. _____

2. List one way this Proverb will strengthen your daily daily walk with God. _____

3. Write out one action step you will take in your life based on today's reading. *I will* _____

Request
-Devote four minutes to talking with God.

Lesson 245

Relax
    -Spend one minute meditating on the words of Colossians 1:16-17. *"All things were created through Him and for Him. He is before all things, and in Him all things hold together."*

Read
    -Devote four minutes to reading 2nd Corinthians 3.

Reflect
    -Take three minutes to think through the following questions.
1. Read verse 3—Why does Paul tell these Christians they are a letter of Christ?
2. Reflect on the following statement—*You may be the only Bible your friends ever read!*
3. In what ways is Jesus superior to Moses?
4. What are some differences between the Old Law and the New Testament as described by Paul in this chapter?

Record
    -Spend three minutes answering the following questions.
1. Write down one question or observation you have over 2nd Corinthians 3. _____
2. List one way today's reading will strengthen your relationship with Jesus. _____
3. Write out one action step you will take in your life based on your study of 2nd Corinthians 3. ***I will*** ___

Request
    -Devote four minutes to talking with God.

Lesson 246

Relax
 -Spend one minute in a quiet place meditating on the following words. *"God has the power to make you completely adequate!" 2nd Corinthians 3:4*

Read
 -Devote four minutes to reading 2nd Corinthians 4.

Reflect
 -Take three minutes to think through the following questions.
   1. What enabled Paul to not lose heart during the trials he faced in his ministry?
   2. What treasure is Paul speaking of in verse 7?
   3. How would you describe the mission of Satan based on what you read in this chapter?
   4. Is there a connection between believing in Christ & talking about Christ? Can you say the words Paul wrote in verse 13—*"I believed, therefore I spoke."*

Record
 -Spend three minutes answering the following questions.
   1. Write down one question or observation you have over this chapter. _____

   2. List one way this chapter will help you not lose heart as a servant of Jesus. _____

   3. Write out one action step this chapter has motivated you to take in your life. *I will* _____

Request
 -Devote four minutes to talking with your Father.

Lesson 247

Relax
    -Spend one minute meditating on the following words. *"To walk with Him is to walk in light. "The Lord is my light and my salvation—whom shall I fear? Psalm 27:1*

Read
    -Devote four minutes to reading Proverbs 16.

Reflect
    -Take three minutes to think through the following questions.
1. Which verse from Proverbs 16 challenges you the most spiritually?
2. Read verse 25—How can one know if they are going the right way in life?
3. How does God feel about those who are proud and arrogant?
4. How important are wisdom and understanding to a Christian? *Are you seeking these things?*

Record
    -Spend three minutes answering the following questions.
1. Write down one question or observation you have over this Proverb. _____

2. List one lesson you learned from this chapter that will help you become more like Jesus. _____

3. Write out one action step you will take in your life based on today's reading. *I will* _____

Request
    -Devote four minutes to talking with God.

Lesson 248

Relax
-Spend one minute in a quiet place talking with God. *Share with Him any struggles you are facing at work, within your family, or in your daily spiritual walk.*

Read
-Devote four minutes to reading 2nd Corinthians 5.

Reflect
-Take three minutes to think through the following questions.
1. How should Christians view death?
2. Reflect on this statement—*What God doesn't fix in this life, He will replace in the next!*
3. Is your focus on the temporal or the eternal?
4. Read verse 7—How can a Christian walk by faith?
5. What motivated Paul to share the gospel with the lost? *How do you feel realizing Christ has left His ministry in your hands?*

Record
-Spend three minutes answering the following questions.
1. How will this chapter help you become a more effective *ambassador* for Christ? _____
   _____
2. Write down one question or observation you have over this chapter. _____
   _____
3. List one action step 2nd Corinthians 5 has motivated you to take in your life. *I will* _____
   _____

Request
-Devote four minutes to talking with God.

Lesson 249

Relax
-Spend one minute meditating on the words of 2nd Corinthians 6:16b. *"We (you) are the temple of the living God."*

Read
-Devote four minutes to reading 2nd Corinthians 6.

Reflect
-Take three minutes to think through the following questions.
1. What are the three greatest commitments a person will make in their life?
2. Are you completely committed to these commitments?
3. How committed was Paul to Jesus Christ? *How committed are you to Jesus Christ?*
4. Read verse 1—How can someone receive God's grace in vain?

Record
-Spend three minutes answering the following questions.
1. Write down one question or observation you have over this chapter. _____
2. Explain the significance of verses 14-18 to you as a Christian. _____
3. Write down one action step this chapter has motivated you to take in your life. *I will* _____

Request
-Devote four minutes to talking with God.

Lesson 250

Relax
-Spend one minute meditating on the following physical description of heaven from Revelation 21:21. *"And the street of the city was pure gold, like transparent glass."*

Read
-Devote four minutes to reading Proverbs 17.

Reflect
-Take three minutes to think through the following questions.
1. Which verse in Proverbs 17 provides you the greatest challenge in your Christian walk?
2. What are some important qualities you look for in a friend?
3. Read verse 17—Based on this verse, are you a good friend to others?
4. What is greater in the world today—A love for God or a love for sin? *Which is greater in your life?*

Record
-Spend three minutes answering the following questions.
1. Write down one question or observation you have over this Proverb. _____

2. List one way this chapter will strengthen your love for God. _____

3. Write out one action step you will take in your life based on today's reading. *I will* _____

Request
-Devote four minutes to talking with your Father.

Lesson 251

Relax
-Spend one minute meditating on some promises God makes to us as His children through His word. *Which promise provides you the greatest motivation to live for God?*

Read
-Devote four minutes to reading 2nd Corinthians 7.

Reflect
-Take three minutes to think through the following questions.
1. What scheme is Satan using on you to try and get you to turn away from God?
2. Reflect on the following statement—*One of Satan's most effective strategies against Christians is to make them too busy for God!*
3. Read verse 2—Do you have room in your heart for God?
4. What are the differences between Godly sorrow and worldly sorrow?

Record
-Spend three minutes answering the following questions.
1. Write down one question or observation you have over 2nd Corinthians 7. _____
2. List one reason why you should have confidence as a Christian. _____
3. Write out one action step this chapter has motivated you to take in your life. *I will* _____

Request
-Devote four minutes to talking with God.

Lesson 252

Relax
    -Spend one minute meditating on the love Jesus has for you. *"Though He (Jesus) was rich, yet for your sake He became poor, so that you through His poverty might become rich."*

Read
    -Devote four minute to reading 2nd Corinthians 8.

Reflect
    -Take three minutes to think through the following questions.
1. What is the greatest physical gift you have ever received?
2. Reflect on the following statement—*Jesus left the glory of heaven to give you the opportunity to be in heaven!*
3. What is the greatest test of a person's love? (Read John 15:13)
4. How would you describe the giving of the Macedonians?

Record
    -Spend three minutes answering the following questions.
1. Write down one question or observation you have over 2nd Corinthians 8. _____
2. How will this chapter help you in your giving? _____
3. Write out one action step you will take in your life based on today's study. *I will* _____

Request
    -Devote four minutes to talking with your Father.

Lesson 253

Relax
   -Spend one minute meditating on the words of Proverbs 18:10. *"The name of the Lord is a strong tower; the righteous runs into it and is safe."*

Read
   -Devote four minutes to reading Proverbs 18.

Reflect
   -Take three minutes to think through the following questions.
   1. What are some different names ascribed to God in the Bible? *Father, Jehovah, Almighty, I Am.....*
   2. Which name of God brings you the most comfort in your life?
   3. What is the connection between pride/arrogance and destruction?
   4. Read verse 24—Why will a man of too many friends come to ruins?

Record
   -Spend three minutes answering the following questions.
   1. Write down one question or observation you have over this Proverb. _____

   2. List one lesson you learned about the marriage relationship from this Proverb. _____

   3. Write out one action step you will take in your life based on today's reading. *I will* _____

Request
   -Devote four minutes to talking with your Father.

Lesson 254

Relax
-Spend one minute meditating on the following quote. *Don't worry about what you can't do for God. Just do what you can! (Mark 14:9)*

Read
-Devote four minutes to reading 2nd Corinthians 9.

Reflect
-Take three minutes to think through the following questions.
1. What is the most important lesson you learned from this chapter?
2. How would you describe God's giving? *What is the ultimate gift God has given humanity?*
3. Read verse 7—What do you learn about giving from this verse?
4. Read and reflect on the story of the widow in Mark 12:41-44.

Record
-Spend three minutes answering the following questions.
1. Write down one question or observation you have over this chapter. _____

2. List one way 2nd Corinthians 9 will strengthen your desire to give your best to God. _____

3. Write out one action step you will take in your life based on today's reading. *I will* _____

Request
-Devote four minutes to talking with God.

Lesson 255

Relax
-Spend one minute meditating on the words of Jesus in Matthew 11:28. *"Come to me, all who are weary and heavy laden and I will give you rest."*

Read
-Devote four minutes to reading 2nd Corinthians 10.

Reflect
-Take three minutes to think through the following questions.
1. What are the differences between having confidence and being arrogant?
2. Does God want us to be confident as Christians?
3. In what should you be confident in as a Christian?
4. How would you describe the war we are in as Christians based on what you read in this chapter?
5. Read verse 5—How can a person take their thoughts captive?

Record
-Spend three minutes answering the following questions.
1. Write out one question or observation you have over this chapter. _____

2. In your opinion, what is the most effective way to share the gospel message? _____

3. Write out one action step this chapter has motivated you to take in your life. *I will* _____

Request
-Devote four minutes to talking with your Father.

Lesson 256

Relax
-Spend one minute meditating on the following quote while thinking of the challenges you face as a Christian. *Great moments are born from great opportunities!*

Read
-Devote four minutes to reading Proverbs 19.

Reflect
-Take three minutes to think through the following questions.
1. How does God feel about those who tell lies? *Is it ever OK in the eyes of God to tell a lie?*
2. How many times is the sin of lying condemned in this Proverb?
3. Is there such a thing as absolute truth? *Where can a person go to find absolute truth? (John 17:17)*
4. Read Proverbs 19:2b—How does constantly being in a hurry impact our effectiveness for Christ?

Record
-Spend three minutes answering the following questions.
1. Write down one question or observation you have over this Proverb. _____

2. List one way this chapter will help you become a more effective servant for God. _____

3. Write out one action step Proverbs 19 has motivated you to take in your life. *I will* _____

Request
-Devote four minutes to talking with God.

Lesson 257

Relax
-Spend one minute meditating on the following words from 2nd Timothy 3:16. *"All Scripture is inspired by God and profitable for teaching, for reproof, for correction, for training in righteousness."*

Read
-Devote four minutes to reading 2nd Corinthians 11.

Reflect
-Take three minutes to think through the following questions.
1. What does the phrase *"All Scripture is God breathed"* mean to you as a Christian?
2. How would you describe the responsibility we have as Christians towards God's word?
3. Were the Corinthians taking a stand for God's word? *Are you taking a stand for God's word?*
4. What did you learn about the disguises of Satan in this chapter?

Record
-Spend three minutes answering the following questions.
1. Write down one question or observation you have from this chapter. _____

2. List one way this chapter will help you get prepared for the attacks of Satan. _____

3. Write out one action step you will take in your life based on today's reading. *I will* _____

Request
-Devote four minutes to talking with God.

Lesson 258

Relax
-Spend one minute meditating on the words of Paul in 2nd Corinthians 12:10. *"I am well content with weaknesses, with insults, with distresses, with persecutions, with difficulties, for Christ's sake; for when I am weak, then I am strong."*

Read
-Devote four minutes to reading 2nd Corinthians 12.

Reflect
-Take three minutes to think through the following questions.
1. Why was Paul given a thorn in the flesh?
2. What important lessons did Paul learn as a result of this thorn?
3. Reflect on the words of Philippians 4:13—*"I can do all things through Christ who strengthens me."*
4. What would it be like to get an actual glimpse of heaven and hell while living on earth?

Record
-Spend three minutes answering the following questions.
1. Write out one question or observation you have over 2nd Corinthians 12. _____

2. Read verse 20—Which one of these sins do you Struggle with the most in you spiritual life? _____

3. Write out one action step this chapter has motivated you to take in your life. *I will* _____

Request
-Devote four minutes to talking with God.

Lesson 259

Relax
-Spend one minute meditating on the following words. *You don't have to wonder about your salvation—you can know with complete confidence you are saved! 1st John 5:13*

Read
-Devote four minutes to reading 2nd Corinthians 13.

Reflect
-Take three minutes to think through the following questions.
1. What is the importance of examining ourselves as Christians?
2. How can we test ourselves as Christians? *What does Paul tell the Corinthians to look for in verse 5?*
3. In verses 9 and 11 Paul writes about being made complete. *How can you accomplish this goal of being made complete in your life?*
4. Whose weakness is Paul referring to in verse 4?

Record
-Spend three minutes answering the following questions.
1. Write down one question or observation you have over this chapter. _____
2. In your opinion, which verse from this chapter would have the greatest impact on someone who was lost? _____
3. Write out one action step you will take in your life based on today's reading. *I will* _____

Request
-Devote four minutes to talking with God.

Lesson 260

Relax
    -Spend one minute meditating on the words of 2nd Corinthians 13:11. *Christianity is the best life to live in part because of the blessings we receive in this life!*

Read
    -Devote four minutes to reading Proverbs 20.

Reflect
    -Take three minutes to think through the following questions.
1. What is the greatest advice a parent or friend has ever given you?
2. What are some blessings associated with walking in integrity?
3. Which verse in this Proverb is most challenging for you to live by?
4. In your opinion, what is the greatest need in the church today? (Read verse 15)

Record
    -Spend three minutes answering the following questions.
1. Write down one question or observation you have over this Proverb. _____

2. List one way this chapter will help you in your relationships with your friends and family. _____

3. Write out one action step you will take in your life based on today's reading. *I will* _____

Request
    -Devote four minutes to talking with your Father.

Lesson 261

Relax
    -Spend one minute meditating on the words of James 1:3 *"knowing that the testing of your faith produces patience"* (NKJV).

Read
    -Devote four minutes to reading James chapter 1.

Reflect
    -Take three minutes to think through the following questions.
1. Why should we count it all joy when we fall into various trials?
2. What are the sources of temptation?
3. What does it mean to be like a man observing his natural face in a mirror?
4. What is pure and undefiled religion?

Record
    -Spend three minutes answering the following questions.
1. Write down one question or observation you have over James chapter 1._____
_____

2. List one way this chapter can help you become a more faithful Christian.
_____
_____

3. Write out one action step you will take in your life based on today's reading. ***I will***
_____
_____

Reflect
    -Devote four minutes to talking with God in prayer.

Lesson 262

Relax

-Spend one minute meditating on the words of James 2:5 *"which He promised to those who love Him."*

Read

-Devote four minutes to reading James chapter 2.

Reflect

-Take three minutes to think through the following questions.
1. How do we show partiality?
2. What is the difference between faith and works?
3. When is judgment without mercy?
4. How was Abraham justified?

Record

-Spend three minutes answering the following questions.
1. Write down one question or observation you have over James chapter 2. _____
_____

2. List one way this chapter can help you become a more faithful Christian. _____
_____

3. Write out one action step you will take in your life based on today's reading. *I will* _____
_____

Reflect

-Devote four minutes to talking to God

Lesson 263

Relax
    -Spend one minute meditating on the words of James 3:6 regarding the tongue.

Read
    -Devote four minutes to reading James chapter 3.

Reflect
    -Take three minutes to think through the following questions.
1. Why does the writer use the example of a bit and rudder?
2. Why is the tongue described as a fire?
3. What does it mean to ask can a spring bring forth fresh water and bitter water?
4. How does the writer describe wisdom from above?

Record
    -Spend three minutes answering the following questions.
1. Write down one question or observation you have over James chapter 3. _____
_____

2. List one way this chapter can help you become a stronger Christian. _____
_____

3. Write out one action step you will take in your life based on today's reading. *I will* _____
_____

Reflect
    -Devote four minutes to talking to the Lord.

Lesson 264

Relax
-Spend one minute meditating on the words found in James 4 that inform us that God resists the proud but gives grace to the humble (v. 6).

Read
-Devote four minutes to reading James chapter 4.

Reflect
-Take three minutes to think through the following questions.
1. What is considered enmity with God?
2. How can we submit ourselves to God?
3. What is required by mankind in order that God will lift us up?
4. How is life described regarding longevity in James 4?

Record
-Spend three minutes answering the following questions.
1. Write down one question or observation you have over James chapter 4. _____
_____

2. List one way this chapter can help you become a more zealous Christian. _____
_____

3. Write out one action step you will take in your life based on today's reading. *I will* _____
_____

Reflect
-Devote four minutes to talking to God

Lesson 265

Relax
    -Spend one minute meditating on the words of James 5 regarding being patient until the coming of the Lord. (v. 7-8).

Read
    -Devote four minutes to reading James chapter 5.

Reflect
    -Take three minutes to think through the following questions.
        1. How can money affect our relationship with God?
        2. Why should we not grumble against one another?
        3. What does it mean to let your "Yes" be "Yes," and your "No," "No"?
        4. What avails much?

Record
    -Spend three minutes answering the following questions.
        1. Write down one question or observation you have over James chapter 5. _____

        2. List one way this chapter can help you become a more passionate Christian. _____

        3. Write out one action step you will take in your life based on today's reading. *I will* _____

Reflect
    -Devote four minutes to talking to Jehovah.

Lesson 266

Relax
-Spend one minute meditating on the words of Philemon 1:4 "I thank my God, making mention of you always in my prayers" (NKJV). Who do you mention in your prayers?

Read
-Devote four minutes to reading Philemon chapter 1.

Reflect
-Take three minutes to think through the following questions.
1. Why was thankful to God for Philemon?
2. Who was Onesimus?
3. Why did Paul send Onesimus back?
4. Are we slaves today?

Record
-Spend three minutes answering the following questions.
1. Write down one question or observation you have over Philemon chapter 1._____

2. List one way this chapter can help you become a more faithful servant. _____

3. Write out one action step you will take in your life based on today's reading. *I will* _____

Reflect
-Devote four minutes to talking to God.

Lesson 267

Relax
  -Spend one minute meditating on the following quote while reflecting on your priorities. *"Count as lost each day you have not used in loving God."*

Read
  -Devote four minutes to reading Ephesians 5.

Reflect
  -Take three minutes to think through the following questions.
  1. What are some motivators Paul describes in verses 1-14 to devoting one's life to morality?
  2. What does the early part of this chapter say concerning the verbal filth so common in our day?
  3. How does the Christian find his/her greatest freedom and fulfillment?
  4. What relationship described in verses 22-33 serves as the foundation for building every relationship?

Record
  -Spend three minutes answering the following questions.
  1. Write down one question or observation you have over this chapter. _____
  _____
  2. List one lesson you learned from this chapter that will help you become a better spouse now or in the future. _____
  3. Write out one action step Ephesians 5 has motivated you to take in your life. *I will* _____
  _____

Request
  -Devote four minutes to talking with God.

Lesson 268

Relax
   -Spend one minute meditating on the following words. *"For the word of God is living and active and sharper than any two edged sword...able to judge the thoughts...of the heart."*

Read
   -Devote four minutes to reading Ephesians 6.

Reflect
   -Take three minutes to think through the following questions.
   1. How would you describe the nature of the Christian's conflict?
   2. What does the armor tell us about the nature of Satan's attack?
   3. How many weapons of offense does a Christian have to use against Satan?
   4. Which part of the Christian armor do you need to strengthen in your life?

Record
   -Spend three minutes answering the following questions.
   1. Write down one question or observation you have over this chapter. _____
   2. List one lesson you learned from Ephesians 6 that will help you overcome Satan. _____
   3. Write out one action step you will take in your life based on today's reading. *I will* _____

Request
   -Devote four minutes to talking with God.

Lesson 269

Relax
   -Spend one minute meditating on the following words.
   "Nothing can give us so great relief in the trials and sorrows of life as a loving relationship with God."

Read
   -Devote four minutes to reading Proverbs 23.

Reflect
   -Take three minutes to think through the following questions.
1. What are some common temptations that are mentioned in this Proverb?
2. How would you describe the nature of temptations based on what you read in this chapter?
3. What is the most challenging temptation you are currently fighting?

Record
   -Spend three minutes answering the following questions.
1. Write out one question or observation you have over this chapter. _____

2. What advice from this Proverb will help you overcome temptations? _____

3. Write out one action step you will take in your life based on today's reading. *I will* _____

Request
   -Devote four minutes to talking with God asking for strength to overcome temptation.

Lesson 270

Relax
-Spend one minute meditating on the follow words. *"God is everywhere, in all places, and there is no spot where we cannot draw near to Him."*

Read
-Devote four minutes to reading 1st Peter 1.

Reflect
-Take three minutes to think through the following questions.
1. What do you learn about God's word from this chapter?
2. Who was Peter writing this letter to? *What are some blessings revealed in this chapter that are only found in Christ?*
3. How valuable is faith in action? *How valuable is your faith?*
4. How long was the redemption of humanity through the sacrifice of Jesus part of God's plan?

Record
-Spend three minutes answering the following questions.
1. Write out one question or observation you have over this chapter. _____

2. What did you learn about Christian living from verses 13-16? _____

3. Write out one action step you will take in your life based on today's reading. *I will* _____

Request
-Devote four minutes to talking with God.

Lesson 271

Relax
    -Spend one minute meditating on the following words. *"The grass withers and the flower falls off, but the word of the Lord endures forever."*

Read
    -Devote four minutes to reading 1st Peter 2.

Reflect
    -Take three minutes to think through the following questions.
1. How would you describe Christianity based on what you read in this chapter?
2. Read verses 5 and 9—Which one of these phrases used to describe Christians is most meaningful to you?
3. Are Christians supposed to suffer? (Read 2nd Timothy 3:12) *Are you suffering for Christ?*
4. Why does God allow Christians to suffer?

Record
    -Spend three minutes answering the following questions.
1. Write out one question or observation you have over 1st Peter 2. _____
2. What did you learn about Jesus in this chapter that will help you follow in His footsteps? _____
3. Write out one action step you will take in your life based on today's reading. *I will* _____

Request
    -Devote four minutes to talking with God.

Lesson 272

Relax
-Spend one minute meditating on the following words. *"The greatest glory we can give to God is to distrust our own strength and to commit ourselves wholly to His safekeeping."*

Read
-Devote four minutes to reading Proverbs 24.

Reflect
-Take three minutes to think through the following questions.
1. What is the only treasure humans possess that will endure forever?
2. Read verse 13—How important is wisdom in the eyes of God?
3. What is the connection between wisdom and your soul?
4. Are you wise in the eyes of God? *Are you devoting more time storing up physical treasures or heavenly treasures?*

Record
-Spend three minutes answering the following questions.
1. Write down one question or observation you have over this Proverb. _____
2. Which verse impacted you the most from today's reading? _____
3. Write out one action step this chapter has motivated you to take in your life. *I will* _____

Request
-Devote four minutes to talking with God.

Lesson 273

Relax
-Spend one minute meditating on the following words. *"We need only to recognize that God is intimately present with us to address ourselves to Him every moment."*

Read
-Devote four minutes to reading 1st Peter 3.

Reflect
-Take three minutes to think through the following questions.
1. How would you describe the marriage relationship based on what you read in 1st Peter 3?
2. What is one God-given responsibility for the woman and the man in the marriage relationship?
3. Read verse 12—Is this verse in reference to this life or in the future?
4. Are you able to fulfill verse 15 in your life?
5. Read verse 21—What does baptism do for a person?

Record
-Spend three minutes answering the following questions.
1. Write down one question or observation you have over this chapter. _____

2. How will 1st Peter 3 help you in your marriage relationship now or in the future? _____

3. Write out one action step you will take in your life based on today's reading. *I will* _____

Request
-Devote four minutes to talking with God.

Lesson 274

Relax
-Spend one minute meditating on the following words. *"To accomplish great things, we must not only act but also dream; not only plan, but also believe."*

Read
-Devote four minutes to reading 1st Peter 4.

Reflect
-Take three minutes to think through the following questions.
1. Why were these Christians facing persecution from the Gentiles?
2. What are some strategies revealed in verses 7-11 that can help you endure trials and persecutions?
3. Read verse 16—What are some blessings that come as a result of suffering for Christ?
4. How do you envision the judgment day based on what you read in this chapter?

Record
-Spend three minutes answering the following questions.
1. Write out one question or observation you have over 1st Peter 4. _____

2. How has this chapter increased your passion to minister to others? _____

3. Write out one action step you will take in your life based on today's reading. *I will* _____

Request
-Devote four minutes to talking with God.

Lesson 275

Relax
-Spend one minute meditating on the following words.
"Cast all your anxiety on God, because He cares for you."

Read
-Devote four minutes to reading 1st Peter 5.

Reflect
-Take three minutes to think through the following questions.
1. In your opinion, what is Satan's most effective weapon he uses on society?
2. What is Satan's greatest weapon he is using to attack you?
3. What are some responsibilities elders must fulfill and qualities elders must possess according to Peter in this chapter?
4. What is the most encouraging lesson you learned from this chapter?

Record
-Spend three minutes answering the following question.
1. Write out one question or observation you have over 1st Peter 5. _____

2. List one strategy you learned from this chapter that will help you overcome Satan. _____

3. Write out one action step this chapter has motivated you to take in your life. *I will* _____

Request
-Devote four minutes to talking with your Father.

Lesson 276

Relax
-Spend one minute meditating on the following words. *"Our heart must be knit in love to God, and our communion with Him so close as to compel us to run to Him at every moment."*

Read
-Devote four minutes to reading Proverbs 25.

Reflect
-Take three minutes to think through the following questions.
1. What is the best advice you've been given that you failed to follow?
2. In your opinion, what is the best piece of advice found in this Proverb?
3. What are some things people possess that cannot be found in looks alone?
4. When adversity comes into your life, how do you handle it?

Record
-Spend three minutes answering the following questions.
1. Write down one question or observation you have over this Proverb. _____
2. What is the most important quality a person needs in order to be a productive Christian? Explain. _____
3. Write out one action step you will take in your life based on today's reading. *I will* _____

Request
-Devote four minutes to talking with God.

Lesson 277

Relax
-Spend one minute meditating on the following question.
"What would life be if we had no courage to attempt anything?"

Read
-Devote four minutes to reading 2nd Peter 1.

Reflect
-Take three minutes to think through the following questions.
1. Why should we be confident as Christians?
2. What is the greatest promise God has given you in His word?
3. How can this chapter help you grow as a Christian?
4. Which trait from verses 5-7 do you need to strengthen in your life?
5. What is the source of all the prophesies found in the pages of the Bible?

Record
-Spend three minutes answering the following questions.
1. Write down one question or observation you have over this chapter. _____

2. How can this chapter help you convince someone the Bible is from God? _____

3. Write out one action step this chapter has motivated you to take in your life. *I will* _____

Request
-Devote four minutes to talking with God.

Lesson 278

Relax
-Spend one minute meditating on the following words while Reflecting on your mission as a Christian. *"The greatest joy in life comes with trying to help others."*

Read
-Devote four minutes to reading 2nd Peter 2.

Reflect
-Take three minutes to think through the following questions.
1. Is there a problem with false teachers in religion today?
2. How can you detect if someone is a false teacher?
3. What did you learn from this chapter that opposes the teaching of "once saved, always saved"?
4. How would you describe God based on what you read about Him in 2nd Peter 2?

Record
-Spend three minutes answering the following questions.
1. Write down one question or observation you have over this chapter. _____

2. List one way this chapter will help you become a more passionate minister for God. _____

3. Write out one action step you will take in your life based on today's reading. *I will* _____

Request
-Devote four minutes to talking with your Father.

Lesson 279

Relax
-Spend one minute meditating on the following words. *"The Lord is......patient toward you, not wishing for any to perish, But for all to come to repentance."*

Read
-Devote four minutes to reading 2nd Peter 3.

Reflect
-Take three minutes to think through the following questions.
1. Why did Peter write this second letter to these Christians?
2. What did you learn about the second coming of Jesus from this chapter?
3. What impact should the coming judgment day have on the lives of Christians?
4. How can a Christian grow in the "grace and knowledge" of Jesus?

Record
-Spend three minutes answering the following questions.
1. Write down one question or observation you have over this chapter. _____

2. List one time when God has demonstrated his patience towards you. _____

3. Write out one action step you will take in your life based on today's reading. *I will* _____

Request
-Devote four minutes to talking with God.

Lesson 280

Relax

-Spend one minute meditating on the following words.
"Too often we underestimate the power of a touch, smile, kind word, a listening ear, an honest compliment or the smallest act of caring, all which have the potential to turn a life around."

Read

-Devote four minutes to reading Proverbs 26.

Reflect

-Take three minutes to think through the following questions.
1. What are the most important relationships in your life?
2. In your opinion, what is the secret to building successful, healthy relationships with others?
3. Is it easier to build relationships or destroy them?
4. What are some things revealed in this Proverb that will quickly destroy any relationship?

Record

-Spend three minutes answering the following questions.
1. Write down one question or observation you have over this Proverb. _____

2. How will this chapter help you strengthen your relationship with God? _____

3. Write out one action step you will take in your life based on today's reading. *I will* _____

Request

-Devote four minutes to talking with your Father.

Lesson 281

Relax
-Spend one minute meditating on the following words. *"He rescued us from the domain of darkness, and transferred us to the kingdom of His beloved Son, in whom we have redemption of sins."*

Read
-Devote four minutes to reading Colossians 1.

Reflect
-Take three minutes to think through the following questions.
1. How would you describe the faith of the Christians at Colossae?
2. Is Paul's prayer in this chapter focused on spiritual matters or physical matters?
3. What is the relationship between Christ and the church?
4. Jesus was fully man and fully God! *How does this chapter explain Jesus' role as both God and man?*

Record
-Spend three minutes answering the following questions.
1. Write down one question or observation you have over this chapter. _____

2. If you could perform one act of kindness that would make the world a better place what would it be? ___

3. Write out one action step this chapter has motivated you to take in your life. *I will* _____

Request
-Devote four minutes to talking with God.

Lesson 282

Relax
-Spend one minute meditating on the following words. *"Failure is the opportunity to begin again with more information."*

Read
-Devote four minutes to reading Colossians 2.

Reflect
-Take three minutes to think through the following questions.
1. What advice does Paul issue in verse 6? *How is your walk God?*
2. Why is having a daily walk with God vital to Christians?
3. What does Paul warn these Christians to look out for in this chapter?
4. If Paul were to write you a personal letter, what would he warn you to avoid?

Record
-Spend three minutes answering the following questions.
1. Write out one question or observation you have over this chapter. _____

2. List two blessings Jesus has brought into your life as as a Christian. _____

3. Write out one action step this chapter has motivated you to take in your life. *I will* _____

Request
-Devote four minutes to talking with God.

Lesson 283

Relax
    -Spend one minute meditating on the judgment scene painted in Colossians 3:4. *"When Christ who is our life appears, then you also will appear with Him in glory."*

Read
    -Devote four minutes to reading Colossians 3.

Reflect
    -Take three minutes to think through the following questions.
1. Fill in the following statement: _____ is life.
2. What would your spouse, children, parents or best friend say your life was about?
3. Read verse 2. What are some things above we can focus on? *How do you fulfill this verse in your life?*
4. Which temptation listed in verses 5, 8 and 9 do you struggle with the most?

Record
    -Spend three minutes answering the following questions.
1. Write out one question or observation you have over Colossians 3. _____
2. Describe God's picture of a home as revealed in verses 18-21. _____
3. Write out one action step you will take in your life based on today's reading. *I will* _____

Request
    -Devote four minutes to talking with God.

Lesson 284

Relax
-Spend one minute in a quiet place talking with God. *Share with Him one struggle you are currently going through and one goal you have recently achieved.*

Read
-Devote four minutes to reading Colossians 4.

Reflect
-Take three minutes to think through the following questions.
1. What does Paul ask these Christians to pray for in verses 2-4?
2. When was the last time you asked God to open you a door to share the gospel with someone?
3. Read verse 4—What should our attitude be towards those who are lost?
4. How would you describe the individual people Paul mentions in this chapter?

Record
-Spend three minutes answering the following questions.
1. Write down one question or observation you have have over this chapter. _____
_____
2. In your opinion, what was Paul's greatest quality as a minister? _____
_____
3. Write out one action step you will take in your life as a result of today's study. *I will* _____
_____

Request
-Devote four minutes to talking with your Father.

Lesson 285

Relax
-Spend one minute meditating on the following words. *"God is everywhere, in all places, and there is no spot where we cannot draw near to Him."*

Read
-Devote four minutes to reading Proverbs 27.

Reflect
-Take three minutes to think through the following questions.
1. What are two goals you are currently working towards achieving in your life?
2. Read verse 17—Are your friends bringing you closer to God or dragging you away from God?
3. What does Solomon say about the heart of man in verse 19?
4. What does God see when He looks into your heart?

Record
-Spend three minutes answering the following questions.
1. Write out one question or observation you have over this Proverb. _____

2. Read verse 1 again. *What is one thing you want to accomplish for God today?* _____

3. Write out one action step you will take in your life as a result of today's study. *I will* _____
_____

Request
-Devote four minutes to talking with God.

Lesson 286

Relax

-Spend one minute meditating on the words of Proverbs 4:23. *"Above all else, guard your heart, for it is the wellspring of life."*

Read

-Devote four minutes to reading Hebrews 1.

Reflect

-Take three minutes to think through the following questions.
1. What is the most important message the writer of Hebrews conveys in this chapter?
2. Write out or high light the things in this chapter that God has done through His Son.
3. How would you explain the relationship between God and Jesus?
4. In what ways is Jesus superior to the angels in heaven?

Record

-Spend three minutes answering the following questions.
1. Write out one question or observation you have over this chapter. _____
2. Which verse in Hebrews 1 contains God the Father's confirmation of Jesus the Son being God? _____
3. Write out one action step this chapter has motivated you to take in your life. *I will* _____

Request

-Devote four minutes to talking with your Father.

Lesson 287

Relax
-Spend one minute meditating on the following words summing up the theme of Hebrews. *Jesus is superior to the angels, to Moses, to the prophets; He is the great high priest.*

Read
-Devote four minutes to reading Hebrews 2.

Reflect
-Take three minutes to think through the following questions.
1. Read Hebrews 2:2; Galatians 3:19 and Acts 7:53. *What did you learn in these verses concerning who the Law/Old Testament was spoken through?*
2. How is the New Testament an improvement over the Old Testament?
3. Read Hebrews 2:3—Through whom was the New Testament initially spoken?
4. Why did Jesus come to earth as a man?

Record
-Spend three minutes answering the following questions.
1. Write out one question or observation you have over this chapter. _____
2. List one way this chapter will help you overcome temptation. _____
3. Write out one action step you will take in your life based on today's reading. *I will* _____

Request
-Devote four minutes to talking with God.

Lesson 288

Relax
-Spend one minute meditating on the help God provides us as His children. *"For in that He Himself has suffered being tempted, He is able to aid those who are tempted."*

Read
-Devote four minutes to reading Hebrews 3.

Reflect
-Take three minutes to think through the following questions.
1. How would you describe the similarities between Jesus and Moses?
2. In what ways was Jesus superior to Moses?
3. Which verse(s) in Hebrews 3 proves the teaching of "once saved always saved" is wrong?
4. What causes people today to harden their hearts towards God?
5. Is your heart open to listen and respond to God?

Record
-Spend three minutes answering the following questions.
1. Write out one question or observation you have over this chapter. _____

2. List one lesson you learned from this chapter that will help you become a better servant. _____

3. Write out one action step Hebrews 3 has motivated you to take in your life. *I will* _____

Request
-Devote four minutes to talking with God.

Lesson 289

Relax
-Spend one minute meditating on these words from Proverbs 28. *"An arrogant man stirs up strife, but he who trusts in the Lord will prosper."*

Read
-Devote four minutes to reading Proverbs 28.

Reflect
-Take three minutes to think through the following questions.
1. Why is it important that we confess our sins to God and to each other?
2. What are some things Christians need to do every day?
3. Read verse 19—What are some empty pursuits people today chase after?
4. Is it easier for you to trust in God or self? *How can we learn to put our complete trust in God?*

Read
-Spend three minutes answering the following questions.
1. Write down one question or observation you have over Proverbs 28. _____

2. Why is it important for the righteous to be bold? ___

3. Write out one action step you will take in your life based on today's reading. *I will* _____

Reflect
-Devote four minutes to talking with God asking specifically for the courage to be *"bold as a lion."*

Lesson 290

Relax
-Spend one minute meditating on the words of Hebrews 4:16. *"Let us draw near with confidence to the throne of grace so that we may receive mercy and find grace to help in time of need."*

Read
-Devote four minutes to reading Hebrews 4 and 5.

Reflect
-Take three minutes to think through the following questions.
1. How does verse 13 make you feel as a Christian? *Does this verse scare you or bring you confidence?*
2. In your opinion, which descriptor of God's word in Hebrews 4:12 is most powerful?
3. Why should we turn to Jesus when we are facing struggles as humans?
4. How is Jesus' relationship as God's Son different than your relationship with God as His child?

Read
-Spend three minutes answering the following questions.
1. Write down one question or observation you have over Hebrews 4 and 5. _____

2. Why is it important for individuals to grow and and mature as Christians? _____

3. Write out one action step these chapters have motivated you to take in your life. *I will* _____

Request
-Devote four minutes to talking with your Father.

Lesson 291

Relax
-Spend one minute meditating on the following words. *"Joy is not the absence of trouble, but the presence of God."*

Read
-Devote four minutes to reading Hebrews 6.

Reflect
-Take three minutes to think through the following questions.
1. How would you describe the hope you have as a Christian?
2. What is one thing you do everyday to grow spiritually?
3. If you had one minute to talk to the world about Jesus Christ, what would you say?
4. How would you describe the purpose of God based on what you read in this chapter?

Record
-Spend three minutes answering the following questions.
1. Write down one question or observation you have over this chapter. _____
2. List some qualities we need as Christians to help us not become sluggish spiritually. _____
3. Write out one action step you will take in your life based on today's reading. *I will* _____

Request
-Devote four minutes to talking with God.

Lesson 292

Relax
-Spend one minute in a quiet place talking with God. *"It was at this time that Jesus went off to the mountain to pray, and He spent the whole night in prayer to God."*

Read
-Devote four minutes to reading Hebrews 7 and 8.

Reflect
-Take three minutes to think through the following questions.
1. How is Melchizedek described in chapter 7? *What is significant about this man?*
2. How is the priesthood of Jesus superior to the priesthood of the Old Testament?
3. Describe the duties of a priest. *What does Jesus offer us as our high priest?*
4. Read Hebrews 8:6-7—In what ways is the New Testament better than the Old Testament?

Record
-Spend three minutes answering the following questions.
1. Write down one question or observation you have over this chapter. _____
2. Read 1st Peter 2:5—As Christians, we are made priests by Jesus. *What is your responsibility as a priest?* _____
3. Write out one action step these chapters have motivated you to take in your life. *I will* _____

Request
-Devote four minutes to talking with God.

Lesson 293

Relax
-Spend one minute meditating on the following words.
"Most men are about as happy as they make up their minds to be."

Read
-Devote four minutes to reading Proverbs 29.

Reflect
-Take three minutes to think through the following questions.
1. What motivated you to become a Christian?
2. What motivates you to stay faithful to Jesus?
3. This Proverb discusses pride, anger, and the love of money. *Which one of these sins do you struggle with the most in your life?*
4. What advice does Solomon give concerning raising children in this Proverb? *Is this advice being followed in your family?*

Record
-Spend three minutes answering the following questions.
1. Write down one question or observation you have over this Proverb. _____

2. List one way this chapter will help you strengthen your daily walk with God. _____

3. Write out one action step you will take in your life based on today's reading. *I will* _____

Request
-Devote four minutes to talking with God.

Lesson 294

Relax
-Spend one minute in a quiet place talking with God thanking Him for the sacrifice He and Jesus made on your behalf.

Read
-Devote four minutes to reading Hebrews 9.

Reflect
-Take three minutes to think through the following questions.
1. What are some distinctions between the Old and the New Testament as described in this chapter?
2. What does all humanity have in common? *We are all sinners in desperate need of a Savior!*
3. What had to transpire in order for the New Testament (Will) to go into effect?
4. What are some promises God has made humanity in this New Will? *How can one receive these promises?*

Record
-Spend three minutes answering the following questions.
1. Write out one question or observation you have over this chapter. _____
2. List one thing Jesus did for you that will help you stay faithful to Him. _____
3. Write out one action step you will take in your life based on today's reading. *I will* _____

Request
-Devote four minutes to talking with God.

Lesson 295

Relax
-Spend one minute meditating on the promises that God has made to you as His child. *Which promise provides you the most motivation in your spiritual life?*

Read
-Devote four minutes to reading Hebrews 10.

Reflect
-Take three minutes to think through the following questions.
1. Why did the Old Testament priests continually make sacrifices for sin?
2. What is significant about Jesus' sacrifice for sin? (see verses 10-12)
3. Why is it important to assemble with other Christians?
4. What is the result of a person rejecting the sacrifice of Jesus?

Record
-Spend three minutes answering the following questions.
1. Write down one question or observation you have over this chapter. _____

2. Why should we be confident as Christians? _____

3. Write out one action step this chapter has motivated you to take in your life. *I will* _____

Request
-Devote four minutes to talking with your Father.

Lesson 296

Relax
- Spend one minute meditating on the following statement from Philippians 3:19. *"God will supply all of your needs through Jesus Christ."*

Read
- Devote four minutes to reading Hebrews 11.

Reflect
- Take three minutes to think through the following questions.
    1. What is God's definition of faith? (See Hebrews 11:1 and James 2:26)
    2. Which 'hero of faith' brings you the most inspiration in your life?
    3. What motivated these individuals to do these amazing things?
    4. Which one of these acts of faith would be most difficult for you to do in your life?

Record
- Spend three minutes answering the following questions.
    1. Write down one question or observation you have over Hebrews 11. _____

    2. List one thing you have been putting off that you need to do for God. _____

    3. Write out one action step you will take in your life based on today's reading. *I will* _____

Request
- Devote four minutes to talking with God.

Lesson 297

Relax
    -Spend one minute meditating on the following words adapted from Ephesians 1:3. *"God has blessed us with every spiritual blessing in Christ."*

Read
    -Devote four minutes to reading Proverbs 30.

Reflect
    -Take three minutes to think through the following questions.
1. What happens to one who adds to the words of God?
2. Who is the best example you know of a Christian and why?
3. Why are the shephanim (rock badgers) considered wise? *What lesson can we learn from these creatures?*
4. Throughout the Proverbs arrogance is condemned. *How can we guard our hearts against arrogance?*

Record
    -Spend three minutes answering the following questions.
1. Write down one question or observation you have over this Proverb. _____

2. If you could change or alter one quality about yourself, what quality would you alter? _____

3. Write out one action step this Proverb has motivated you to take in your life. *I will* _____

Request
    -Devote four minutes to talking with God.

Lesson 298

Relax
   -Spend one minute meditating on the following words from Hebrews 13:8. *"Jesus Christ is the same yesterday and today and forever."*

Read
   -Devote four minutes to reading Hebrews 12.

Reflect
   -Take three minutes to think through the following questions.
   1. Who is the ultimate example of faith?
   2. Why should we fix our eyes on Jesus?
   3. What are some things that can prevent us from running the Christian race?
   4. What causes people today to lose heart? *How can you prevent yourself from losing heart as a Christian?*
   5. Read verse 28—How are we to worship God? *What should motivate us to fulfill this command?*

Record
   -Spend three minutes answering the following questions.
   1. Write down one question or observation you have over this chapter. _____

   2. Why does God discipline His children? _____

   3. Write out one action step this chapter has motivated You to take in your life. *I will* _____

Request
   -Devote four minutes to talking with God.

Lesson 299

Relax
-Spend one minute meditating on the promise Jesus makes to you in Hebrews 13:5. *"I will never desert you, nor will I ever forsake you."*

Read
-Devote four minutes to reading Hebrews 13.

Reflect
-Take three minutes to think through the following questions.
1. What are some practical things found in this chapter one can do to demonstrate their Christianity?
2. What does this chapter teach concerning the marriage relationship?
3. Which verse from this chapter brings you the most confidence as a Christian?
4. Read verse 5 and reflect on the following statement. *"Contentment is learned."*

Record
-Spend three minutes answering the following questions.
1. Write down one observation or question you have over this chapter. _____

2. List one lesson you learned from this chapter that will help you become a better minister to others. ___

3. Write out one action step you will take in your life based on today's reading. *I will* _____

Request
-Devote four minutes to talking with God.

Lesson 300

Relax
    -Spend one minute meditating on the words of Proverbs 16: 16. *"How much better it is to get wisdom than gold! And to get understanding is to be chosen above silver."*

Read
    -Devote four minutes to reading Proverbs 31.

Reflect
    -Take three minutes to think through the following questions.
1. How valuable is an *excellent wife*?
2. What are the virtues an excellent wife will possess?
3. Describe the relationships a virtuous woman has with her family and with others.
4. What responsibilities will a virtuous woman fulfill?
5. Which characteristic do you admire most about the woman described in this Proverb?

Record
    -Spend three minutes answering the following questions.
1. Write down one question or observation you have over this Proverb. _____
2. List one way this chapter will improve your relationships with your family. _____
3. Write out one action step this Proverb has motivated you to take in your life. *I will* _____

Request
    -Devote four minutes to talking with God.

Lesson 301

Relax
-Spend one minute in a quiet place talking with God. *Share any challenges you are currently facing with Him.*

Read
-Devote four minutes to reading Mark 1.

Reflect
-Take three minutes to think through the following questions.
1. How would you describe John?
2. Why were so many drawn to the preaching of John?
3. In what ways did John help prepare the way for Jesus?
4. Why was it necessary for Jesus to overcome the attacks of Satan?
5. Describe the faith of Simon, Andrew, James and John.

Record
-Spend three minutes answering the following questions.
1. Write down one question or observation you have over Mark 1. _____

2. What did you learn about Jesus' personal relationship with God in this chapter? _____

3. Write out one action step this chapter has motivated you to take in your life. *I will* _____

Request
-Devote four minutes to talking with God in a solitary place as Jesus did in verse 35.

Lesson 302

Relax
-Spend one minute meditating on the following words.
Success is not avoiding failure, but picking yourself up each time you fail.

Read
-Devote four minutes to reading Mark 2.

Reflect
-Take three minutes to think through the following questions.
1. Why were so many people coming to see Jesus at this time in His ministry?
2. If you had the power to bring anyone to Christ, who would you choose? *What's stopping you?*
3. What draws people to Jesus today? *What causes people to turn away from Jesus today?*
4. Describe the following people you read about in Mark 2: *Jesus, the paralytic, the four friends, the the crowd, the scribes.*

Record
-Spend three minutes answering the following questions.
1. Do you believe Jesus is more concerned with the paralytics spiritual healing or physical healing? ___
2. List some qualities the four friends possess you need in your life. _____
3. Write out one action step you will take in your life based on today's reading. *I will* _____

Request
-Devote four minutes to talking with God.

Lesson 303

Relax
-Spend one minute meditating on the following words.
"Courage is not defined by those who fought and did not fall, but by those who fought, fell and rose again."

Read
-Devote four minutes to reading Mark 3.

Reflect
-Take three minutes to think through the following questions.
1. Why were the Pharisees on a mission to 'catch' Jesus?
2. Which verse in this chapter proves believing in Jesus as the Son of God is not sufficient to save someone?
3. Why did Jesus command the unclean spirits to not testify as to who He was?
4. What are some different accusations Jesus had to overcome in this chapter?

Record
-Spend three minutes answering the following questions.
1. Write down one question or observation you have over Mark 1. _____

2. List one way this chapter will strengthen your faith in Jesus as God's Son. _____

3. Write out one action step this chapter has motivated you to take in your life. *I will* _____

Request
-Devote four minutes to talking with God.

Lesson 304

Relax

-Spend one minute meditating on the following; *"The only difference between those who threw in the towel and quit and those who use their energy to rebuild and keep going is found in the word hope."*

Read

-Devote four minutes to reading Mark 4.

Reflect

-Take three minutes to think through the following questions.
1. In what ways has Jesus Christ changed your life?
2. Define hope. *Concerning hope, where would you be without Jesus?*
3. Which type of soil in Mark 4 best describes your spiritual heart?
4. Read verse 20—How can you tell if your heart is good?

Record

-Spend three minutes answering the following questions.
1. Write down one question or observation you have over this chapter. _____
2. Why does Jesus compare the kingdom of heaven to a mustard seed? _____
3. Write out one action step this chapter has motivated you to take in your life. *I will* _____

Request

-Devote four minutes to talking with God.

Lesson 305

Relax
-Spend one minute meditating on the *"great things Jesus has done for you and how he has been merciful to you."*

Read
-Devote four minutes to reading Mark 5.

Reflect
-Take three minutes to think through the following questions.
1. What do you learn about Jesus from His interaction with Legion?
2. Why did the people who witnessed this miracle beg Jesus to leave their area?
3. What message did Jesus tell Legion to share with others?
4. In your opinion, who is the greatest story of faith found in this chapter?
5. Are you taking time to tell others what great things Jesus has done for you in your life?

Record
-Spend three minutes answering the following questions.
1. Write down one question or observation you have over this chapter. _____

2. List one way this chapter will strengthen your ability to minister to others. _____

3. Write out one action step this chapter has motivated you to take in your life. *I will* _____

Request
-Devote four minutes to talking with your Father.

Lesson 306

Relax

-Spend one minute in a quiet place talking with God. *Only as we stop in meditation will our hearts be filled with praise for our friends, our past blessings and our God.*

Read

-Devote four minutes to reading Mark 6.

Reflect

-Take three minutes to think through the following questions.
1. Why did the people in Jesus' hometown take offense at Him?
2. What are some challenges a person will face when choosing to follow Jesus?
3. How would you describe the relationship between John and Herod?
4. Which verses in this chapter indicate Jesus' strong desire to have a close relationship with His Father?

Record

-Spend three minutes answering the following questions.
1. Write down one question or observation you have over Mark 6. _____

2. List one characteristic you have in common with the disciples. _____

3. Write out one action step this chapter has motivated you to take in your life. *I will* _____

Request

-Devote four minutes to talking with God. (Mark 6:46)

Lesson 307

Relax
-Spend one minute meditating on the following words in relation to your spiritual walk. *"If we do not learn from the mistakes of history, we are bound to repeat them."*

Read
-Devote four minutes to reading Mark 7.

Reflect
-Take three minutes to think through the following questions.
1. What is the most important lesson Jesus teaches in this chapter?
2. What is the significance of an individual's spiritual heart?
3. How can you know if your heart is good or bad?
4. Read verses 20-23: How can you control your thoughts, and overcome temptation as a Christian?
5. What is the condition of your spiritual heart?

Record
-Spend three minutes answering the following questions.
1. Write down one question or observation you have over this chapter. _____

2. List one lesson you learned from the individuals who were healed by Jesus in Mark 7. _____

3. Write out one action step this chapter has motivated you to take in your life as a Christian. *I will* _____

Request
-Devote four minutes to talking with God.

Lesson 308

Relax
-Spend one minute meditating on the following words of Jesus. *"For what does it profit a man to gain the whole world, and forfeit his soul?"*

Read
-Devote four minutes to reading Mark 8.

Reflect
-Take three minutes to think through the following questions.
1. Jesus is often referred to as the Son of Man in Scripture. *What do we learn about Jesus from this phrase?*
2. Why did Jesus rebuke Peter in verse 33?
3. In your opinion, which verse from this chapter is the heart of Christianity?
4. What is the greatest obstacle you face in striving to build a strong relationship with God?

Record
-Spend three minutes answering the following questions.
1. Write down one question or observation you have over Mark 8. _____
_____
2. In verse 29 Peter described Jesus as the Christ. *How would you describe Jesus to your friends?* _____
_____
3. Write out one action step you will take in your life based on today's reading. *I will* _____
_____

Request
-Devote four minutes to talking with God.

Lesson 309

Relax
-Spend one minute meditating on the following words.
"Perfect resignation to God and His word is a sure way to heaven."

Read
-Devote four minutes to reading Mark 9.

Reflect
-Take three minutes to think through the following questions.
1. What is the significance of the words found in verse 7 for your life?
2. Why does Jesus refer to these people as an "unbelieving generation"?
3. Does this phrase accurately describe the society we live in today? Explain.
4. What are some barriers that prevent people from believing in Jesus as the Messiah?
5. How does Jesus describe hell in this chapter?

Record
-Spend three minutes answering the following questions.
1. Write out one question or observation you have over this chapter. _____

2. What is the secret to greatness in the eyes of Jesus? _____

3. Write down one action step this chapter has motivated you to take in your life. *I will* _____

Request
-Devote four minutes to talking with your Father.

Lesson 310

Relax
   -Spend one minute meditating on the following words. *"Unquestionably, God's kingdom is being advanced by people of little talent doing little jobs for a big God."*

Read
   -Devote four minutes to reading Mark 10.

Reflect
   -Take three minutes to think through the following questions.
   1. How would you describe God's plan for marriage?
   2. What qualities do children possess that we as adults need to emulate?
   3. How would you describe the man Jesus met in verse 17?
   4. Why was it difficult for the disciples to embrace Jesus' teaching on being a servant?

Record
   -Spend three minutes answering the following questions.
   1. Write down one question or observation you have over this chapter. _____
   2. List one thing you can do to strengthen your marriage. _____
   3. Write down one action step you will take in your life based on today's reading. *I will* _____

Request
   -Devote four minutes to talking with God.

Lesson 311

Relax
 -Spend one minute meditating on the words of Isaiah 6:8.
 "I heard the voice of the Lord saying, whom shall I send, and who will go for Us. Then I said, Here am I! Send me."

Read
 -Devote four minutes to reading Mark 11.

Reflect
 -Take three minutes to think through the following questions.
 1. How would you describe the reaction of the people as Jesus entered Jerusalem?
 2. What illustrates the humility of Jesus from this chapter?
 3. Describe the different emotions Jesus experienced as He entered Jerusalem.
 4. In what ways was the teaching of Jesus different from that of the scribes?

Record
 -Spend three minutes answering the following questions.
 1. Write down one question or observation you have over Mark 11. _____
 2. List one way this chapter will strengthen your relationship with Jesus. _____
 3. Write out one action step you will take in your life based on today's reading. *I will* _____

Request
 -Devote four minutes to talking with God.

Lesson 312

Relax
  -Spend one minute meditating on the words of Philippians 4:7. *"And the peace of God, which surpasses all understanding will guard your hearts and minds in Christ Jesus."*

Read
  -Devote four minutes to reading Mark 12.

Reflect
  -Take three minutes to think through the following questions.
  1. Why did Jesus share the parable of the vine growers in verses 1-11?
  2. What was the greatest obstacle preventing the scribes, Pharisees and Sadducees from believing in Jesus?
  3. Read verse 28—What is the importance of love in our relationship with God, and with other people?
  4. Compare your giving to that of the widow in verses 41-44. *Are you giving your best to God?*

Record
  -Spend three minutes answering the following questions.
  1. Write down one question or observation you have over this chapter. _____
  2. How would you describe your current relationship with God? _____
  3. Write out one action step this chapter has motivated you to take in your life. *I will* _____

Request
  -Devote four minutes to talking with God.

Lesson 313

Relax
-Spend one minute meditating on the following words. *At the time of the writing of Paul's letter to the church at Colossae, every person had been given the chance to hear the gospel.*

Read
-Devote four minutes to reading Mark 13.

Reflect
-Take three minutes to think through the following questions.
1. How did the apostles and other Jews feel about the temple in Jerusalem?
2. What did Jesus tell His apostles concerning the buildings in Jerusalem?
3. What is the significance of verse 31 for all of humanity?
4. Think about the coming judgment day; *are you ready for Jesus to come?*

Record
-Spend three minutes answering the following questions.
1. Write down one question or observation you have over this chapter. _____

2. Read verses 10 and 14: What did Jesus say would happen before Jerusalem would be destroyed? ____

3. List one action step this chapter has motivated you To take in your life. *I will* _____

Request
-Devote four minutes to talking with your Father.

Lesson 314

Relax
-Spend one minute meditating on the following words. *"All things were created through Him and for Him. He is before all things and in Him all things hold together."*

Read
-Devote four minutes to reading Mark 14.

Reflect
-Take three minutes to think through the following questions.
1. How would you describe Jesus' emotions while in the garden?
2. What did Jesus encourage His apostles to do while He was in the garden praying? Why?
3. Why were the apostles unable to stand up for Jesus?
4. Read verse 36—Are you living your life to please God or self?
5. How strong is your daily relationship with God?

Record
-Spend three minutes answering the following questions.
1. Write down one question or observation you have over this chapter. _____

2. Describe the suffering Jesus endured in this chapter. _____

3. Write out one action step this chapter has motivated you to take in your life. *I will* _____

Request
-Devote four minutes to talking with God.

Lesson 315

Relax
    -Spend one minute meditating on the following words. *"Live life as if you were going to die tomorrow; learn as if you will live forever."*

Read
    -Devote four minutes to reading Mark 15.

Reflect
    -Take three minutes to think through the following questions.
1. In what ways does the above quote apply to life as a Christian?
2. How would you describe the life of Jesus?
3. Why was Jesus handed over to be crucified? *How does jealousy affect relationships?*
4. How would you describe Pilate based on what you read about him in this chapter?
5. Why did Jesus die on the cross?

Record
    -Spend three minutes answering the following questions.
1. Write down one question or observation you have over this chapter. _____
2. Which person(s) from this chapter best represents you spiritually? Explain. _____
3. Write out one action step you will take in your life based on today's reading. *I will* _____

Request
    -Devote four minutes to talking with God.

Lesson 316

Relax
-Spend one minute meditating on the following words. *"We are made for God and God alone. Our challenge each day is to forsake all, even ourselves, to find our all in Him."*

Read
-Devote four minutes to reading Mark 16.

Reflect
-Take three minutes to think through the following questions.
1. What is the biggest struggle you are currently facing as a Christian?
2. What are some things you want to achieve for God's kingdom in the upcoming year?
3. How would you have reacted if you would have been sitting at the table with the apostles when Jesus appeared to them?
4. What is the significance of the resurrection for you as a Christian?

Record
-Spend three minutes answering the following questions.
1. Write down one question or observation you have over this chapter. _____
2. List two ways you have grown as a Christian this past year. _____
3. Write out one action step this chapter has motivated you to take in your life. *I will* _____

Request
-Devote four minutes to talking with God.

Lesson 317

Relax
-Spend one minute meditating on the following words of Paul in his letter to the Philippians. *"For me to live is Christ and to die is gain."*

Read
-Devote four minutes to reading Philippians 1.

Reflect
-Take three minutes to think through the following questions.
1. Describe Paul's condition has he writes to the Philippians. *How did Paul feel about his situation?*
2. How would you describe Paul's purpose for life based on this chapter?
3. What is the primary purpose of your life?
4. How did Paul feel about the Christians at Philippi?
5. Read verse 27—How can you accomplish Paul's plea in this verse to live in a manner worthy of the gospel? *What is the importance of doing this?*

Record
-Spend three minutes answering the following questions.
1. Write down one question or observation you have over this chapter. _____

2. List one lesson you learned from this chapter that will strengthen your love for Jesus. _____

3. Write out one action step you will take in your life based on today's reading. *I will* _____

Request
-Devote four minutes to talking with God.

Lesson 318

Relax
-Spend one minute meditating on the following quote.
"Peace to the Christian is not the absence of trouble, but the presence of God."

Read
-Devote four minutes to reading Philippians 2.

Reflect
-Take three minutes to think through the following questions.
1. In your opinion, which verse from this chapter best describes the mission of Christians?
2. What is the ultimate expression of Christ's selflessness and obedience?
3. Do you agree with the following statement; *the greatest obstacle you'll face in striving to live for God is the person you see in the mirror everyday.*
4. If we "empty ourselves" what will that mean in practical terms for you?

Record
-Spend three minutes answering the following questions.
1. List three individuals who emptied themselves from Philippians 2. _____
2. Read Philippians 2:2—In striving to be like Christ, which one of these do you struggle with the most? _____
3. Write out one action step you will take in your life based on today's reading. ***I will*** _____

Request
-Devote four minutes to talking with God.

Lesson 319

Relax
  -Spend one minute meditating on the following words. *"The central event in the drama of salvation is an act of complete selfless love."*

Read
  -Devote four minutes to reading Philippians 3.

Reflect
  -Take three minutes to think through the following questions.
  1. In what ways are confidence and arrogance different?
  2. In what should we base our confidence in as Christians?
  3. In your opinion, what is the most powerful statement made by Paul in this chapter?
  4. What is the importance of our example as Christians to those who don't know Christ?

Record
  -Spend three minutes answering the following questions.
  1. List two obstacles Paul had to overcome in his Christian walk. _____

  2. List two obstacles you have overcome as you strive to follow Jesus. _____

  3. Write out one action step this chapter has motivated you to take in your Christian life. *I will* _____

Request
  -Devote four minutes to talking with God.

Lesson 320

Relax
-Spend one minute meditating on Paul's words in Philippians 4:13. *"I can do all things through Christ who strengthens me."*

Read
-Devote four minutes to reading Philippians 4.

Reflect
-Take three minutes to think through the following questions.
1. What impact do quarrels have on the church?
2. At this moment in your life, what is your greatest worry? *What secret does Paul give us to overcome our worries?*
3. What controls your thoughts? *How can you guard your mind as a Christian?*
4. How was Paul able to find contentment in all situations?

Record
-Spend three minutes answering the following questions.
1. Write down one question or observation you have over this chapter. _____
2. Which verse from this chapter brings you the most comfort as a Christian? Why? _____
3. Write out one action step this chapter has motivated you to take in your life. *I will* _____

Request
-Devote four minutes to talking with God.

Lesson 321

Relax
-Spend one minute meditating on the following words.
"Never, never, never, never, never give up. In matters both great and small!"

Read
-Devote four minutes to reading 1 Thessalonians 1 and 2.

Reflect
-Take three minutes to think through the following questions.
1. How would you describe the Thessalonians based on your reading today?
2. In your opinion, what was the greatest compliment Paul bestowed on these Christians?
3. What were some past sins some of the Thessalonians struggled with in their lives?
4. How do you overcome temptation in your daily life as a Christian?

Record
-Spend three minutes answering the following questions.
1. Write down one question or observation you have over these two chapters. _____

2. List one quality Paul possessed as a minister that you need to imitate in your life. _____

3. Write out one action step these chapters have motivated you to take in your life. *I will* _____

Request
-Devote four minutes to talking with your Father.

Lesson 322

Relax
-Spend one minute meditating on the following words.
"Responsibility is the key to greatness."

Read
-Devote four minutes to reading 1 Thessalonians 3 and 4.

Reflect
-Take three minutes to think through the following questions.
1. Why did Paul send Timothy to check on the Christians at Thessalonica?
2. What is the importance of accountability in our Christian walk?
3. Who holds you accountable to stay faithful as a Christian?
4. How should Christians view the second coming of Jesus Christ?
5. For what purpose has God called you as a Christian?

Record
-Spend three minutes answering the following questions.
1. List one responsibility Paul encouraged these Christians to fulfill in their lives. _____

2. List one question or observation you have over these two chapters. _____

3. Write out one action step these chapter have motivated you to take in your life. *I will* _____

Request
-Devote four minutes to talking with God.

Lesson 323

Relax
-Spend one minute meditating on the following words.
"We owe to Scripture the same reverence which we owe to God."

Read
-Devote four minutes to reading 1 Thessalonians 5.

Reflect
-Take three minutes to think through the following questions.
1. What are three important purposes of the church?
2. What did you learn about the mission of the church from this chapter?
3. Read verses 14-22: Which one of these is your greatest strength? *Which one of these areas do you need to strengthen in your Christian life?*
4. What are some practical things you can do in your life to encourage other Christians?

Record
-Spend three minutes answering the following questions.
1. Write down one question or observation you have over this chapter. _____

2. On a scale of 1-10 with 10 being powerful, how would you describe your current prayer life? _____

3. List one action step you will take in your life based on today's reading. *I will* _____

Request
-Devote four minutes to talking with God.

Lesson 324

Relax
-Spend one minute meditating on the following words.
"Examine everything carefully; hold fast to that which is good and abstain from every form of evil."

Read
-Devote four minutes to reading 2 Thessalonians 1.

Reflect
-Take three minutes to think through the following questions.
1. In what areas of their spiritual walk were the Thessalonians excelling?
2. In what ways have you suffered persecution due to your faith in Jesus?
3. Which two groups of people will Jesus punish upon His second coming?
4. Do you simply know facts about God, or do you have a personal relationship with Him?

Record
-Spend three minutes answering the following questions.
1. Write down one question or observation you have over this chapter. _____

2. List one way this chapter will strengthen your relationship with God. _____

3. Write out one action step this chapter has motivated You to take in your life. *I will* _____

Request
-Devote four minutes to talking with God.

Lesson 325

Relax
-Spend one minute meditating on the following words. *"The fruit of the Spirit is love, joy, peace, patience, kindness, goodness, faithfulness, gentleness and self-control."*

Read
-Devote four minutes to reading 2 Thessalonians 2 & 3.

Reflect
-Take three minutes to think through the following questions.
1. What are some promises Christians have from God that are revealed in these chapters?
2. In your opinion, what is the greatest promise God has made you as a Christian?
3. What are some expectations we sometimes have of God that He has not guaranteed us in this life?
4. What are some specific commands Paul gives these Christians in chapter 2?

Record
-Spend three minutes answering the following questions.
1. Write down one question or observation you have over these two chapters. _____

2. List one way you can fulfill the command found in chapter 2 verse 13 of today's reading. _____

3. Write out one action step you will take in your life based on today's reading. *I will* _____

Request
-Devote four minutes to talking with God.

Lesson 326

Relax
-Spend one minute meditating on the following message from God. *"Dear friends, I urge you, as aliens and strangers in the world, to abstain from sinful desires, which war against your soul."*

Read
-Devote four minutes to reading 1 Timothy 1.

Reflect
-Take three minutes to think through the following questions.
1. How would you describe Paul's relationship with Timothy?
2. What should be the goal of our teaching as Christians?
3. What are some different ways you can teach those who are lost?
4. How can a Christian prevent themselves from losing their faith as Hymenaeus and Alexander did?

Record
-Spend three minutes answering the following questions.
1. Write out one question or observation you have over today's reading. _____

2. Which verse in this chapter best captures God's mercy for all humanity? Explain. _____

3. Write out one action step this chapter has motivated you to take in your Christian walk. *I will* _____

Request
-Spend one minute talking with your Father.

Lesson 327

Relax
    -Spend one minute meditating on the following words. *"For I know the plans I have for you, declares the Lord, plans to prosper you.... plans to give you hope and a future."*

Read
    -Devote four minutes to reading 1 Timothy 2.

Reflect
    -Take three minutes to think through the following questions.
1. How does God feel concerning the salvation of humanity?
2. What is a mediator? *In what ways, does Jesus serve as our mediator to God?*
3. Why does God give the command to women found in verses 11 and 12?
4. How can you prevent yourself from being deceived by Satan?

Record
    -Spend three minutes answering the following questions.
1. Write out one question or observation you have over 1 Timothy 2. _____
2. What surprises you most about Christianity? _____
3. List one action step you will take in your life based on today's reading. *I will* _____

Request
    -Devote four minutes to talking with God.

Lesson 328

Relax
-Spend one minute meditating on God's words to Joshua.
"Be strong and of good courage; do not be afraid… for the Lord your God is with you wherever you go."

Read
-Devote four minutes to reading 1 Timothy 3 and 4.

Reflect
-Take three minutes to think through the following questions.
1. Why is God so specific in His requirements for men who serve as elders and deacons?
2. In your opinion, which elder qualification listed in verses 2-6 of chapter 3 is most challenging to fulfill?
3. What religion has fulfilled the apostasy prediction found in chapter 4 verse 3?
4. Do you spend more time building your physical fitness or your spiritual fitness?

Record
-Spend three minutes answering the following questions.
1. Write down one question or observation you have over these two chapters. _____
2. "…in speech, conduct, love, faith and purity show yourself an example of those who believe." *Which one of these things is most difficult for you to be an example in? Explain.* _____
3. Write out one action step these chapters have motivated you to take in your life. *I will* _____

Request
-Devote four minutes to talking with your Father.

Lesson 329

Relax
-Spend one minute meditating on the following words.
Devote your life to the pursuit of righteousness, godliness, faith, love, perseverance and gentleness.

Read
-Devote four minutes to reading 1 Timothy 5 and 6.

Reflect
-Take three minutes to think through the following questions.
1. Is it sinful to compromise? *What are some situations when compromising is sinful?*
2. In what areas of life are you most tempted to compromise God's standards?
3. Why are so many in religion compromising the words of Jesus?
4. Read chapter 6 verse 11: Are you fulfilling Paul's plea to Timothy in your life?

Record
-Spend three minutes answering the following questions.
1. Write down one question or observation you have over these two chapters. _____

2. List one lesson you learned today that will help you strengthen your walk with God. _____

3. Write out one action step you will take in your life based on today's reading. *I will* _____

Request
-Devote four minutes to talking with God.

Lesson 330

Relax
-Spend one minute meditating on the following words. *"God has not given us a spirit of timidity, but of power and love and discipline."*

Read
-Devote four minutes to reading 2 Timothy 1.

Reflect
-Take three minutes to think through the following questions.
1. What do you learn about the gospel from this chapter?
2. How was Paul able to remain faithful through all the struggles he endured?
3. Reflect on this statement; the key to staying devoted to Christ is to know Him.
4. Read verse 12: Do you know Christ the way Paul knew Him?

Record
-Spend three minutes answering the following questions.
1. Write down one question or observation you have over this chapter. _____
2. Who has influenced you the way Eunice and Lois influenced Timothy? _____
3. List one action step you will take in your life based on today's reading. *I will* _____

Request
-Devote four minutes to talking with God.

Lesson 331

Relax
    -Spend one minute meditating on the following words Paul wrote to Timothy. *"For if we endure (with Jesus), we will also reign with Him."*

Read
    -Devote four minutes to reading 2 Timothy 2.

Reflect
    -Take three minutes to think through the following questions.
1. What was Paul's motivation in living for Jesus Christ?
2. Read verse 8—Why does Paul encourage Timothy to remember Jesus? *Are you remembering Jesus in your life?*
3. In your opinion, what is the most important piece of advice Paul offers Timothy in this chapter?
4. Why do so many in the religious world today fail to "accurately handle the word of truth"?
5. Which of the things listed in verse 22 is most challenging for you to pursue? Why?

Record
    -Spend three minutes answering the following questions.
1. Write down one question or observation you have over this chapter. _____

2. Write out one action step this chapter has motivated you to take in your life? *I will* _____

Request
    -Devote four minutes to talking with your Father.

Lesson 332

Relax
-Spend one minute meditating on the following words from 2 Timothy 2:19. *"The Lord knows those who are His."*

Read
-Devote four minutes to reading 2 Timothy 3.

Reflect
-Take three minutes to think through the following questions.
1. How would you describe society today? *Which words or phrases in verses 1-7 fit today's world?*
2. How can a Christian resist the temptation to be like the world?
3. Reflect on verses 16-17—What is the importance of these verses for you and the world?
4. What are some persecutions you have endured as a disciple of Jesus?

Record
-Spend three minutes answering the following questions.
1. Write down one question or observation you have over this chapter. _____

2. What is the most important lesson you learned from 2 Timothy 3? _____

3. Write out one action step this chapter has motivated you to take in your life. *I will* _____

Request
-Devote four minutes talking with God.

Lesson 333

Relax
-Spend one minute meditating on Paul's description of his Christian life. *"I have fought the good fight, I have finished the course, I have kept the faith."*

Read
-Devote four minutes to reading 2 Timothy 4.

Reflect
-Take three minutes to think through the following questions.
1. What motivates you to stay faithful to Jesus?
2. What are some things revealed in this chapter that served as motivation for Paul?
3. Read verse 7—Could you write these things about your life based on how you are currently living?
4. How can you avoid the temptation to turn back to the world as Demas did?

Record
-Spend three minutes answering the following questions.
1. Write down one question or observation you have over this chapter. _____

2. Describe Paul's relationship with God at the time of this writing to Timothy. _____

3. Write out one action step you will take in your life based on today's reading. *I will* _____

Request
-Devote four minutes to talking with your Father.

Lesson 334

Relax
-Spend one minute meditating on the following words. *The Christian journey is a road of learning, trying, failing, trying again; a journey whose end will be glorious for the faithful.*

Read
-Devote four minutes to reading Titus 1.

Reflect
-Take three minutes to think through the following questions.
1. What separates your relationship with God from your other relationships?
2. In what ways does verse 2 affect your relationship with God?
3. Is it easy or difficult for you to trust in God?
4. What are some challenges elders must address in their work?
5. In your opinion, what is the most important quality an elder must possess?

Record
-Spend three minutes answering the following questions.
1. Write down one question or observation you have over this chapter. _____
2. List one quality God possesses that you need to strengthen in your life. _____
3. Write out one action step this chapter has motivated you to take in your life. *I will* _____

Request
-Devote four minutes to talking with God.

Lesson 335

Relax
　　-Spend one minute meditating on the following words. *"You are only as strong as your purpose therefore let us choose reasons to act that are big, bold, righteous and eternal."*

Read
　　-Devote four minutes to reading Titus 2.

Reflect
　　-Take three minutes to think through the following questions.
　　　　1. Who has been the greatest influence on your life as a Christian?
　　　　2. Think about whom you are mentoring for God. *What are some steps you can take to improve your influence?*
　　　　3. How important was mentoring others to Paul?
　　　　4. Read verse 14-15—What are some expectations Paul had of Titus as a fellow Christian? *Are you doing these things in your life?*

Record
　　-Spend three minutes answering the following questions.
　　　　1. Write down one question or observation you have over Titus 2. _____

　　　　2. How would you describe the grace of God? _____

　　　　3. Write down one action step you will take in your life based on today's reading. *I will* _____

Request
　　-Devote four minutes to talking with God.

Lesson 336

Relax
- Spend one minute mediating on the following words. *"You cannot escape the responsibility of tomorrow by evading it today."*

Read
- Devote four minutes to reading Titus 3.

Reflect
- Take three minutes to think through the following questions.
    1. In what ways has God shown you His kindness?
    2. How has God's grace impacted your life?
    3. In what ways can people take God's grace for granted? (See Romans 5:20 – 6:2)
    4. How many times does Paul emphasize the importance of Christians engaging in "good deeds"?
    5. Why does Paul emphasize this to Titus?

Record
- Spend three minutes answering the following questions.
    1. Write down one question or observation you have over this chapter. _____
    2. List some things Paul encourages Titus to avoid in his Christian walk. _____
    3. Write down one action step you will take in your life based on today's reading. *I will* _____

Request
- Devote four minutes to talking with God.

Lesson 337

Relax
   -Spend one minute meditating on the following words. *"The weak can never forgive. Forgiveness is the attribute of the strong."*

Read
   -Devote four minutes to reading Philemon.

Reflect
   -Take three minutes to think through the following questions.
   1. When faced with a moral conflict that demands a decision, how likely are you to go to the Bible for the answer?
   2. What evidences are there in your life that the power of the gospel changes people just like it did Philemon and Onesimus?
   3. Historically, what would have been an acceptable thing for Philemon to do with Onesimus, a run away slave?

Record
   -Spend three minutes answering the following questions.
   1. Write down the words and phrases used in this letter that show the type of relationship Paul had with Onesimus. _____

   2. List one action step you will take in your life based on today's reading. *I will* _____

Request
   -Devote four minutes to talking with your Father.

Lesson 338

Relax
-Spend one minute meditating on your current relationship with God.

Read
-Devote four minutes to reading 1 John 1.

Reflect
-Take three minutes to think through the following questions.
1. In your opinion, what is the most powerful statement John makes in this chapter?
2. What gives John credibility as a writer?
3. In this chapter John describes God as light. If you had to describe God with only one word, what would it be? Why?
4. Why is it important for Christians to be confident in their salvation?
5. Which verse from this chapter brings you the most confidence as a Christian?

Record
-Spend three minutes answering the following questions.
1. Write down one question or observation you have over this chapter. _____
2. Based on this chapter, describe a Christians relationship towards sin. _____
3. Write out one action step this chapter has motivated you to take in your life. *I will* _____

Request
-Devote four minutes to talking with God.

Lesson 339

Relax
   -Spend one minute meditating on the following words. *"The world is passing away and also its lust; but the one who does the will of God lives forever."*

Read
   -Devote four minutes to reading 1 John 2.

Reflect
   -Take three minutes to think through the following questions.
   1. How can an individual know whether or not they are saved?
   2. Read verse 16—What temptations are you facing in each of these areas?
   3. Why did John write this letter to these Christians?
   4. In what ways does this chapter motivate you to live for Christ?

Record
   -Spend three minutes answering the following questions.
   1. Write down one question or observation you have over this chapter. _____

   2. In this chapter John mentions antichrists. How does John describe antichrists? (verse 22) _____

   3. Write out one action step you will take in your life based on today's reading. *I will* _____

Request
   -Devote four minutes to talking with your Father.

Lesson 340

Relax
-Spend one minute meditating on the following words. *There is nothing stronger than the heart of a volunteer.*

Read
-Devote four minutes to reading 1 John 3.

Reflect
-Take three minutes to think through the following questions.
1. How would you define sin? *How does John describe sin in this chapter?*
2. What should a Christian's relationship be towards the world?
3. Are your thoughts, attitudes and actions becoming more like Christ everyday?
4. Reflect on the following quote in relation to your life: *The devil has been vanquished, but he has not vanished.*

Record
-Spend three minutes answering the following questions.
1. Write down one question or observation you have over 1 John 3. _____

2. How does John describe the purpose for which Jesus left heaven to come to earth? _____

3. Write out one action step this chapter has motivated you to take in your life. *I will* _____

Request
-Devote four minutes talking with God.

Lesson 341

Relax
-Spend one minute meditating on the following words. *"For God so loved the world that He gave His only begotten Son that whoever believes in Him shall not perish, but have eternal life."*

Read
-Devote four minutes to reading 1 John 4.

Reflect
-Take three minutes to think through the following questions.
1. What are some things that motivate you to follow God?
2. In what way does love motivate you to follow God?
3. What are some differences between God's view of love and the world's view of love?
4. How can we learn to love others the way God has loved us?

Record
-Spend three minutes answering the following questions.
1. Write down one question or observation you have over this chapter. _____

2. What is the importance of Christians showing love to other people? _____

3. Write out one action step this chapter has motivated you to take in your life. *I will* _____

Request
-Devote four minutes to talking with God.

Lesson 342

Relax
  -Spend one minute meditating on the following words.
  "Within the covers of the Bible are all the answers to all the problems that face us today, if we'd only look."

Read
  -Devote four minutes to reading 1 John 5.

Reflect
  -Take three minutes to think through the following questions.
  1. Why did John write in this chapter that the commands of God are not burdensome?
  2. For what reason did John write this letter to these Christians?
  3. Why is it important for Christians to know they are saved?
  4. What are the three most important things in your life? *Does how your living right now reflect this?*

Record
  -Spend three minutes answering the following questions.
  1. Write out down one question or observation you have over this chapter. _____
  _____
  2. In your opinion, what is the greatest earthly blessing that comes from a relationship with God? _____
  _____
  3. Write out one action step this chapter has motivated you to take in your life. *I will* _____
  _____

Request
  -Devote four minutes to talking with God.

Lesson 343

Relax
 -Spend one minute meditating on the following words. *The man who refused to forgive destroys the bridge over which he himself must cross.*

Read
 -Devote four minutes to reading 2 & 3 John.

Reflect
 -Take three minutes to think through the following questions.
 1. What is the importance of truth and love in the Christian walk?
 2. How can Christians take a stand for truth in a loving manner?
 3. What is the result of neglecting one of these vital areas?
 4. What is the secret to maintaining an ongoing relationship with both God and Jesus?

Record
 -Spend three minutes answering the following questions.
 1. Write down one question or observation you have over these two chapters. _____

 2. What important lesson did you learn from Diotrephes in 3 John? _____

 3. Write out one action step you will take in your life based on today's reading. *I will* _____

Request
 -Devote four minutes to talking with your Father.

Lesson 344

Relax

-Spend one minute meditating on the following words. *"To read the Bible is to take a trip to the fair land where the spirit is strengthened and faith renewed."*

Read

-Devote four minutes to reading Revelation 1.

Reflect

-Take three minutes to think through the following questions.
1. What are some different ways Jesus describes Himself in Revelation 1?
2. In your opinion, which description of Jesus was most important to the seven churches?
3. Revelation is a message to stay faithful to Jesus in all circumstances: *When do you find it most difficult to stay faithful to Jesus?*
4. What are some things listed in this chapter that Jesus brings to us as Christians?

Record

-Spend three minutes answering the following questions.
1. Write down one question or observation you have over this chapter. _____
2. Why did Jesus and John have credibility with these churches? _____
3. Write out one action step you will take in your life based on today's reading. *I will* _____

Request

-Devote four minutes to talking with God.

Lesson 345

Relax
 -Spend one minute meditating on the following words of Jesus. *"Be faithful until death and I will give you the crown of life."*

Read
 -Devote four minutes to reading Revelation 2.

Reflect
 -Take three minutes to think through the following questions.
 1. What is the most important lesson you learn from each of these churches in chapter 2?
 2. Which one of these churches best describes your current walk with God?
 3. What do we learn about Jesus' expectations for His church from this chapter?
 4. How can the Lord's church today follow in the footsteps of the church at Smyrna?

Record
 -Spend three minutes answering the following questions.
 1. Write down one question or observation you have over this chapter. _____

 2. List one thing you learned about Jesus from Revelation 2. _____

 3. Write out one action step this chapter has motivated you to take in your life. *I will* _____

Request
 -Devote four minutes to talking with God.

Lesson 346

Relax
-Spend one minute in a quiet place reflecting on your spiritual walk. *Ask God for help with one temptation you are currently facing.*

Read
-Devote four minutes to reading Revelation 3.

Reflect
-Take three minutes to think through the following questions.
1. What was preventing the church at Sardis from finding approval with Jesus?
2. How did Jesus end His message to each of these seven churches in chapters 2 and 3?
3. Why does Jesus condemn being lukewarm over being cold in our spiritual life?
4. In what ways had Satan tricked the church at Laodicea? Does Satan do this today?

Record
-Spend three minutes answering the following questions.
1. Write out one question or observation you have over this chapter. _____

2. What is the most important lesson you learned from Revelation 3. _____

3. Write out one action this chapter has motivated you to take in your spiritual life. *I will* _____

Request
-Devote four minutes to talking with God.

Lesson 347

Relax
    -Spend one minute meditating on the following words.
    "Holy, holy, holy is the Lord God, the Almighty, who was and who is and who is to come."

Read
    -Devote four minutes to reading Revelation 4.

Reflect
    -Take three minutes to think through the following questions.
1. What do you think John was thinking as he saw heaven opened up before him?
2. Would you live differently if you were given a glimpse of heaven? What about hell?
3. How were the words and images in this chapter helpful to John and other persecuted Christians?
4. In your opinion, what one word in this chapter best describes the nature of God? Why?

Record
    -Spend three minutes answering the following questions.
1. Write down one question or observation you have over Revelation 4. _____
2. List one lesson you learn about worship from the scenes in this chapter. _____
3. Write out one action step you will take in your life based on today's reading. *I will* _____

Request
    -Devote four minutes to talking with your Father.

Lesson 348

Relax
-Spend one minute meditating on the following words. *The task ahead of us is never as great as the Power behind us.*

Read
-Devote four minutes to reading Revelation 5.

Reflect
-Take three minutes to think through the following questions.
1. What do we learn from the failed search in verse 3?
2. Why did John weep when no one was found to be worthy to open the book?
3. How is Jesus described in this chapter? *Why is Jesus described both as a lion and a lamb?*
4. What are some things revealed in this chapter that Jesus brings us as Christians?

Record
-Spend three minutes answering the following questions.
1. Write out one question or observation you have over Revelation 5. _____

2. What is the most important lesson you learn about Jesus in this chapter? _____

3. Write out one action step this chapter has motivated you to take in your life. *I will* _____

Request
-Devote four minutes to talking with God.

Lesson 349

Relax
-Spend one minute meditating on the following words. *One today is worth two tomorrows. Make good use of today.*

Read
-Devote four minutes to reading Revelation 6.

Reflect
-Take three minutes to think through the following questions.
1. Why were the things revealed when the fifth seal was broken both devastating and satisfying?
2. Reflect on your Christian walk in relation to what Paul wrote Timothy in 2 Timothy 3:12.
3. What unlikely image is seen when the sixth seal was broken? *Why is Jesus seen in this manner?*
4. How would you describe the events that were seen when the first four seals were broken? *Were these images helpful or harmful to these Christians?*

Record
-Spend three minutes answering the following questions.
1. Write down one question or observation you have over Revelation 6. _____

2. What did you learn about the judgment day from verses 15-17? _____

3. Write out one action step this chapter has motivated you to take in your spiritual life. *I will* _____

Request
-Devote four minutes to talking with God.

Lesson 350

Relax
 -Spend one minute meditating on the following words. *"A religion that gives nothing, costs nothing, and suffers nothing is worth nothing."*

Read
 -Devote four minutes to reading Revelation 7.

Reflect
 -Take three minutes to think through the following questions.
 1. What is the importance of verse 3 for the faithful Christian?
 2. What does this verse tell us about God's relationship with those who are faithful?
 3. What did it cost many of these Christians to follow Jesus?
 4. Which verse(s) in this chapter brings you the most hope as a Christian? Why?

Record
 -Spend three minutes answering the following questions.
 1. Write down one question or observation you have over this chapter. _____

 2. On a scale of 1-10 with 10 being 'great', where would you rank your current relationship with God? _____

 3. List one action step this chapter has motivated you to take in your life. *I will* _____

Request
 -Devote four minutes to talking with God.

Lesson 351

Relax
-Spend one minute reflecting on the following promise from God found in Hebrews 13. *"I will never desert you, nor will I will ever forsake you."*

Read
-Devote four minutes to reading Revelation 8.

Reflect
-Take three minutes to think through the following questions.
1. In your opinion, what is the greatest promise God gives us in His word?
2. In what way do God's promises impact your Christian walk?
3. Reflect on the following verses in the NT that speak to the power of prayer; *Matthew 7:7-8; James 5:16; John 16:23.*
4. How would you describe the events of this chapter?

Record
-Spend three minutes answering the following questions.
1. Write down one question or observation you have over this chapter. _____
2. Write a sentence that best describes your current prayer life. _____
3. List one action step this chapter has motivated you to take in your life. *I will* _____

Request
-Devote four minutes to talking with God.

Lesson 352

Relax
-Spend one minute meditating on the following words introducing Psalm 23. *"The Lord is my Shepherd I shall not want."*

Read
-Devote four minutes to reading Revelation 9.

Reflect
-Take three minutes to think through the following questions.
1. What is the greatest obstacle you have overcome as a Christian?
2. How has overcoming this challenge helped you in your Christian walk?
3. Reflect on 2 Peter 3:9—What is God's ultimate goal for humanity?
4. Define repentance. Why is it difficult for us as humans to repent?

Record
-Spend three minutes answering the following questions.
1. Write down one question or observation you have over this chapter. _____
2. How would you describe the events revealed to John in this chapter? _____
3. List one action step you will take in your life based on today's reading. *I will* _____

Request
-Devote four minutes to talking with God.

Lesson 353

Relax
    -Spend one minute meditating on the following words.
    "Success is the ability to go from one failure to another with no loss of enthusiasm."

Read
    -Devote four minutes to reading Revelation 10.

Reflect
    -Take three minutes to think through the following questions.
1. Why was John commanded to not write the things revealed to him by the 7 peals of thunder?
2. What does this tell us about God's plan for humanity? *(See Hebrews 11:1)*
3. In your opinion, what does the little book in this chapter represent?
4. How is the message of the gospel both sweet and Bitter at the same time? *(Read Romans 1:16 and 2 Thessalonians 1:8)*

Record
    -Spend three minutes answering the following questions.
1. Write down one question or observation you have over this chapter. _____
2. What was John commanded to do at the conclusion of this chapter? _____
3. List one action step this chapter has motivated you to take in your life. *I will* _____

Request
    -Devote four minutes to talking with God in prayer.

Lesson 354

Relax

-Spend one minute meditating on the following words. *"Shut out all your past except that which will help you weather your tomorrows."*

Read

-Devote four minutes to reading Revelation 11.

Reflect

-Take three minutes to think through the following questions.
1. In what way do the two messengers in this chapter represent the mission of the church?
2. In your opinion, who are these two men based on the descriptions listed in verse 6?
3. What is the overall theme presented in this chapter?
4. Which verse in this chapter provides you the most motivation to stay faithful to God?

Record

-Spend three minutes answering the following questions.
1. Write down one question or observation you have over this chapter. _____

2. List the different names that are ascribed to the righteous in this chapter. _____

3. List one action step you will take in your life based on today's reading. *I will* _____

Request

-Devote four minutes to talking with God in prayer.

Lesson 355

Relax
    -Spend one minute meditating on the following words in relation to your Christian walk. *"Motivation is what gets you started. Habit is what keeps you going."*

Read
    -Devote four minutes to reading Revelation 12.

Reflect
    -Take three minutes to think through the following questions.
1. How would you describe Satan's mission after reading this chapter?
2. Reflect on some instances in Scripture where Satan attempted to destroy Jesus?
3. Read Matthew 16:18—What does this verse along with Revelation 12 reveal about Satan's plan to eliminate the church?
4. What is the secret to overcoming Satan's attacks?

Record
    -Spend three minutes answering the following questions.
1. Write down one question or observation you have over this chapter. _____
2. Which verse in Revelation 12 brings you the most comfort as a Christian? _____
3. Write down one action step you will take in your life based on today's reading. *I will* _____

Request
    -Devote four minutes to talking with God.

Lesson 356

Relax
-Spend one minute meditating on the following words. *"There is nothing more important than personal salvation, nothing."*

Read
-Devote four minutes to reading Revelation 13.

Reflect
-Take three minutes to think through the following questions.
1. What does the beast out of the sea represent? (Read Daniel 4:4-7)
2. How would you describe the beast that is introduced in verse 11? *This beast represents false prophets.*
3. What lessons do you learn about the workings of Satan from this chapter?
4. How do you respond to the struggles that come your way as a Christian?

Record
-Spend three minutes answering the following questions.
1. Write down one question or observation you have over this chapter. _____
2. In your opinion, what was the biggest obstacle these Christians were facing? _____
3. Write out one action step this chapter has motivated you to take in your Christian walk. *I will*_____
_____

Request
-Devote four minutes to talking with God.

Lesson 357

Relax
-Spend one minute meditating on the following beatitude from Revelation 14. *"Blessed are the dead who die in the Lord."*

Read
-Devote four minutes to reading Revelation 14.

Reflect
-Take three minutes to think through the following questions.
1. Which quality of the faithful Christians listed in verses 4-5 is most challenging you to achieve?
2. Read 2 Corinthians 11:2—How is the relationship between Jesus and the church described in this verse?
3. Based on Revelation 14:13, how should Christians view death?
4. Do you fear death?

Record
-Spend three minutes answering the following questions.
1. Write down one question or observation you have over this chapter. _____

2. What thoughts do you have over the judgment scene described in verses 19-20? _____

3. List one action step you will take in your life based on today's reading. *I will* _____

Request
-Devote four minutes to talking with God.

Lesson 358

Relax
-Spend one minute meditating on the following words. *"I have been driven many times to my knees by the overwhelming conviction that I had nowhere else to go."*

Read
-Devote four minutes to reading Revelation 15

Reflect
-Take three minutes to think through the following questions.
1. If you had to describe God with only one word, what word would you choose?
2. Define holiness. How does this word describe all of God's attributes?
3. Read 1 Peter 1:13-16—How would you explain the expectations God has for His children?
4. What is the significance of Revelation 15:8? *A day is coming when repentance is no longer possible!*

Record
-Spend three minutes answering the following questions.
1. Write down one question or observation you have over this chapter. _____
_____
2. How would you describe the wrath of God? _____
_____
3. List one action step this chapter has motivated you to take in your life. *I will* _____
_____

Request
-Devote four minutes to talking with God in prayer.

Lesson 359

Relax
    -Spend one minute meditating on the following words.
    "Nothing sets a person so much out of the devil's reach as humility."

Read
    -Devote four minutes to reading Revelation 16.

Reflect
    -Take three minutes to think through the following questions.
    1. How can a Christian build a powerful, personal, daily walk with God?
    2. What are some things a Christian should do everyday?
    3. How would you describe the connection between repentance and humility?
    4. Why were the individuals described in this chapter unwilling to repent?

Record
    -Spend three minutes answering the following questions.
    1. Write down one question or observation you have over Revelation 16. _____

    2. What is the most important lesson you learned about God from this chapter? _____

    3. Write out one action step this chapter has motivated you to take in your life. *I will* _____

Request
    -Devote four minutes to talking with God.

Lesson 360

Relax
-Spend one minute meditating on the following words in relation to your Christian walk. *"Two roads diverged in a wood, and I—I took the one less traveled by, and that has made all the difference."*

Read
-Devote four minutes to reading Revelation 17.

Reflect
-Take three minutes to think through the following questions.
1. How would you describe the judgment day based on what you have read in Revelation?
2. Reflect on your current spiritual condition: Is your name written in the book of life?
3. How would you describe the world's view toward sin?
4. What is God's view toward sin?

Record
-Spend three minutes answering the following questions.
1. Write down one question or observation you have over this chapter. _____
2. List one obstacle you are currently facing in your daily walk with God. _____
3. Write out one action step you will take in your life based on today's reading. *I will* _____

Request
-Devote four minutes to talking with your Father.

Lesson 361

Relax
-Spend one minute meditating on the following question. *At this point in your life, what spiritual blessing are you most grateful for?*

Read
-Devote four minutes to reading Revelation 18.

Reflect
-Take three minutes to think through the following questions.
1. How would you describe the attitude of the kings, merchants and sailors in this chapter?
2. What are some things listed here we are tempted to place above God?
3. Why was the harlot (Rome) being punished?
4. Which of the following temptations do you struggle with the most in your life; lust for power, money or possessions?

Record
-Spend three minutes answering the following questions.
1. Write out one question or observation you have over this chapter. _____

2. List one important lesson you learn from Revelation 18. _____

3. Write out one action step this chapter has motivated you to take in your spiritual life. *I will* _____

Request
-Devote four minutes to talking with God.

Lesson 362

Relax
-Spend one minute meditating on the following words.
"Happiness is when what you think, what you say, and what you do are in harmony."

Read
-Devote four minutes to reading Revelation 19.

Reflect
-Take three minutes to think through the following questions.
1. What are some different words used in verses 1-6 to describe God?
2. Which verse in the chapter teaches that angels should not be worshipped?
3. Why is the marriage relationship used to describe Jesus' relationship with the church?
4. Why is Jesus described by the name Word of God in verse 13? *How has Jesus fulfilled this title?*

Record
-Spend three minutes answering the following questions.
1. Write down one question or observation you have over Revelation 19. _____

2. In your opinion, which descriptor of Jesus in verses 11-16 is most powerful. _____

3. Write out one action step this chapter has motivated you to take in your life. *I will* _____

Request
-Devote four minutes to talking with God.

Lesson 363

Relax
-Spend one minute meditating on the following words. *The Lord will watch over your coming and going both now and forevermore.*

Read
-Devote four minutes to reading Revelation 20.

Reflect
-Take three minutes to think through the following questions.
1. What is the most important lesson you learned from this chapter?
2. In what ways is Satan powerless over you as a Christian?
3. What verse from this chapter best describes the awesome power of God?
4. Read verse 12—How are we as individuals authors of our eternal fate?
5. Based on how you are living right now, is your name written in the book of life?

Record
-Spend three minutes answering the following questions.
1. Write down one question or observation you have over this chapter. _____

2. List one action step you will take in your life based on today's reading. *I will* _____

Request
-Devote four minutes to talking with God.

Lesson 364

Relax
-Spend one minute meditating on the following promise from God to the faithful. *"...He will dwell among them, and they shall be His people, and God Himself will be among them."*

Read
-Devote four minutes to reading Revelation 21.

Reflect
-Take three minutes to think through the following questions.
1. In your opinion, what will be the greatest aspect of being in heaven?
2. What are some different words and phrases used to describe God in this chapter?
3. Read verse 5—How does this verse affect your thoughts on heaven?
4. What is the most important lesson you learned from reading this chapter?

Record
-Spend three minutes answering the following questions.
1. Write down one question or observation you have over this chapter. _____

2. Read verse 7—In what ways are the blessings of God conditional? _____

3. List one action step this chapter has motivated you to take in your life. *I will* _____

Request
-Devote four minutes to talking with God.

Lesson 365

Relax
- Spend one minute meditating on the words of the prophet Isaiah in Isaiah 43:1-5.

Read
- Devote four minutes to reading Revelation 22.

Reflect
- Take three minutes to think through the following questions.
    1. How many times does Jesus say "He is coming quickly" in this chapter?
    2. How does this repeated promise from Jesus make you feel?
    3. Which phrases describing Jesus in this chapter are also used in description of God in chapter 21?
    4. If the Lord returned today, would you spend eternity in heaven?

Record
- Spend three minutes answering the following questions.
    1. Write down one question or observation you have over this chapter. _____
    2. What facts did you learn about the coming judgmen day from this chapter? _____
    3. List one action step you will take in your life based on today's reading. *I will* _____

Request
- Devote four minutes to talking with God in prayer.

**Written by Phillip Johnson
May 12, 2006**

© Copyright Focus Press, Inc.
and Phillip Johnson 2011